W9-BSE-682
This is
Sophia Mateo's
(NO TOUCH please)
(if not)

SCIENCE

LEVEL FIVE

SECOND EDITION

Colorado Springs, Colorado

SCIENCE

LEVEL FIVE

SECOND EDITION

Student

Vice President for Purposeful Design Publications
Steven Babbitt

Assistant Director for Textbook Development
Don Hulin

Designers
Ana Brodie
Claire Coleman
Susanna Garmany

Editors
Jennifer Bollinger
Gary Brohmer
Barbara Carpenter
Cheryl Chiapperino
Janice Giles
Ellen Johnson
Macki Jones
Kim Pettit
Nancy Sutton
Lorraine Wadman
Lisa Wood

Second edition 2014

Printed in the United States of America
27 26 25 24 23 22 5 6 7 8 9 10

Elementary Science, Level Five – Student Edition
Purposeful Design Elementary Science series
ISBN 978-1-58331-534-7, Catalog #20051

Purposeful Design Publications is the publishing division of the Association of Christian Schools International (ACSI) and is committed to the ministry of Christian school education, to enable Christian educators and schools worldwide to effectively prepare students for life. As the publisher of textbooks, trade books, and other educational resources within ACSI, Purposeful Design Publications strives to produce biblically sound materials that reflect Christian scholarship and stewardship and that address the identified needs of Christian schools around the world.

References to books, computer software, and other ancillary resources in this series are not endorsements by ACSI. These materials were selected to provide teachers with additional resources appropriate to the concepts being taught and to promote student understanding and enjoyment.

Purposeful Design Publications
A Division of ACSI
731 Chapel Hills Drive • Colorado Springs, CO 80920
800/367-0798 • www.purposefuldesign.org

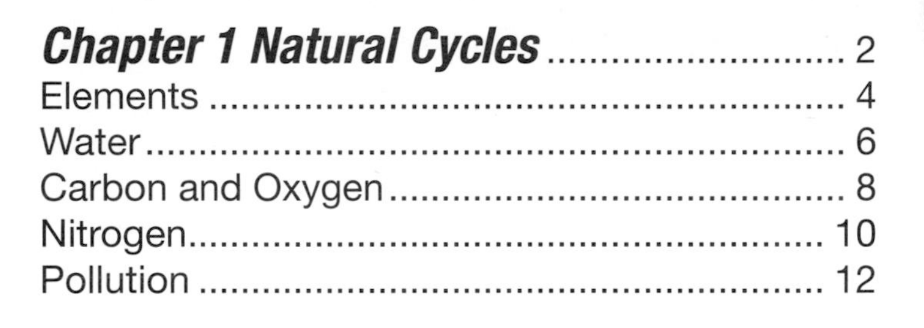

LIFE SCIENCE: CYCLES

UNIT 1

UNIT 2

PHYSICAL SCIENCE: TRANSFORMATIONS

EARTH AND SPACE SCIENCE: PREDICTABILITY

HUMAN BODY: BALANCE

RESOURCES

Development Team

First Edition

Vice President of Academic Affairs
Derek Keenan, EdD

Director of Publishing Services
Steven Babbitt

Assistant Director for Textbook Development
Don Hulin

Authors
Chard Berndt, Pat Blackburn, Sandra Burner, Sue Drake, Dr. Diane Foley, Amy Gruetzmacher, Jacqueline Lauriat, Susan Lovelace, Vince Palmer, Elda Robinson, Pam Van der Werff

Editorial Team
Cynthia Behr, Cheryl Chiapperino, Suzanne Clark, John Conaway, Mary Corcoran, Kristi Crosson, Anita Gordon, David Hill, Stephen Johnson, Macki Jones, JoAnn Keenan, Christy Krenek, Sheri Leasure, Wayne Lowe, Zach Moore, Frieda Nossaman, Vanessa Rough, Kara Underwood, Lisa Wood

Design Team
Lindsay Driscoll, Susanna Garmany, Scot McDonald, Kris Orr, Dan Schultz, Sarah Schultz, Chris Tschamler, Shelley Webb

Consultants
Cheryl Blackmon, PhD
Briarwood Christian School

Bob Burtch, EdD
Wheaton College

Don DeYoung, PhD
Grace College

Ruth Ebeling, MS
Biola University

Ray Gates, MS
Cornerstone University

Jerry Johnson, MS
Corban College

Virginia Johnson, PhD
Biola University

James Van Eaton, PhD
Liberty University

Purposeful Design Publications is grateful to Briarwood Christian School in Birmingham, Alabama, for the contributions they made to the original content of the Purposeful Design Science series.

SCIENTIFIC INQUIRY

Scientific inquiry is a way to help you better understand the world God has created. It involves asking questions, making hypotheses, conducting tests, and analyzing results. New discoveries may direct you to make new hypotheses and to conduct additional tests. Communicating and defending your conclusions is another important step in this ongoing process.

Through this process of exploration, you will develop the ability to formulate pertinent questions, conduct appropriate investigations and tests, and generate reasonable explanations for your findings.

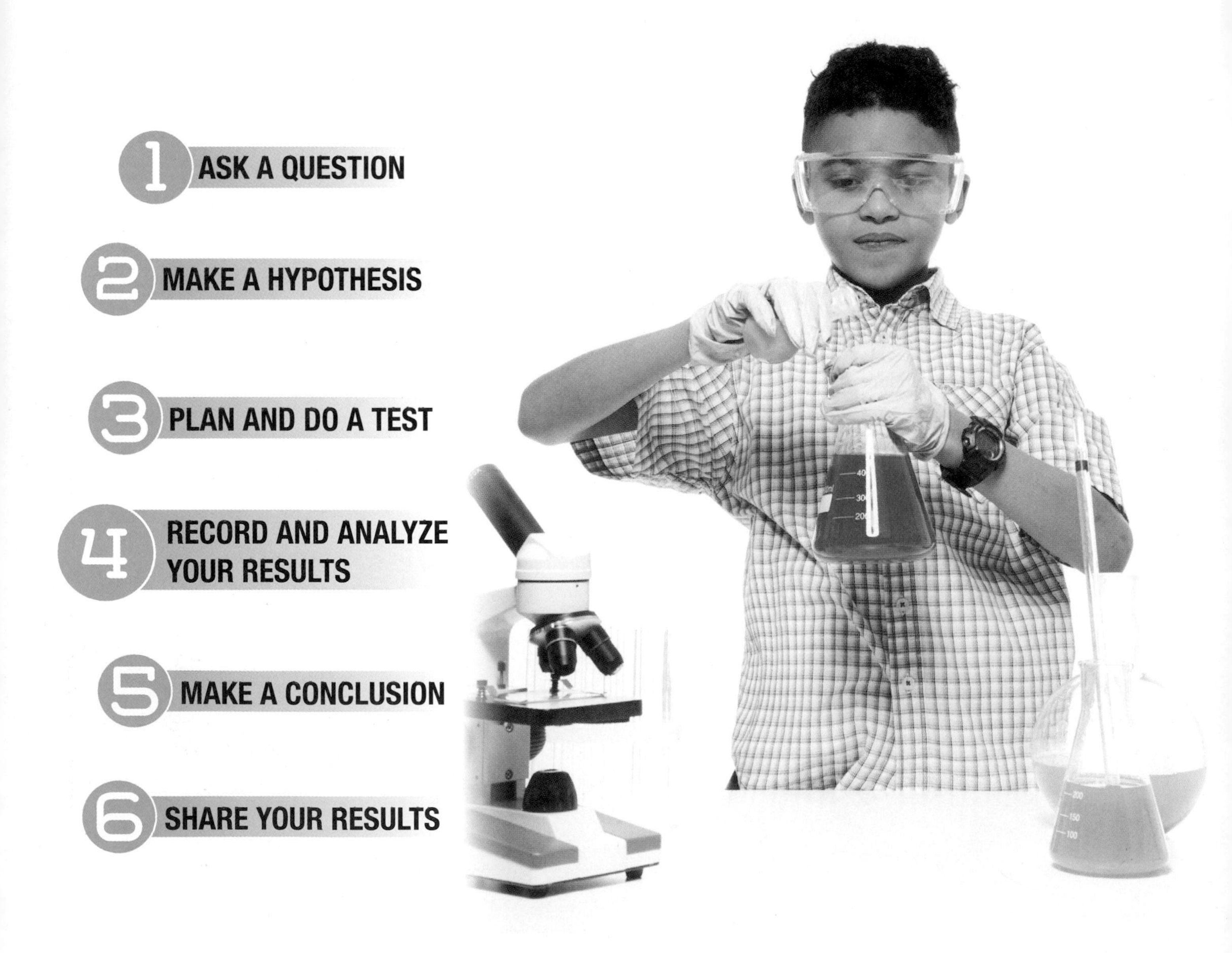

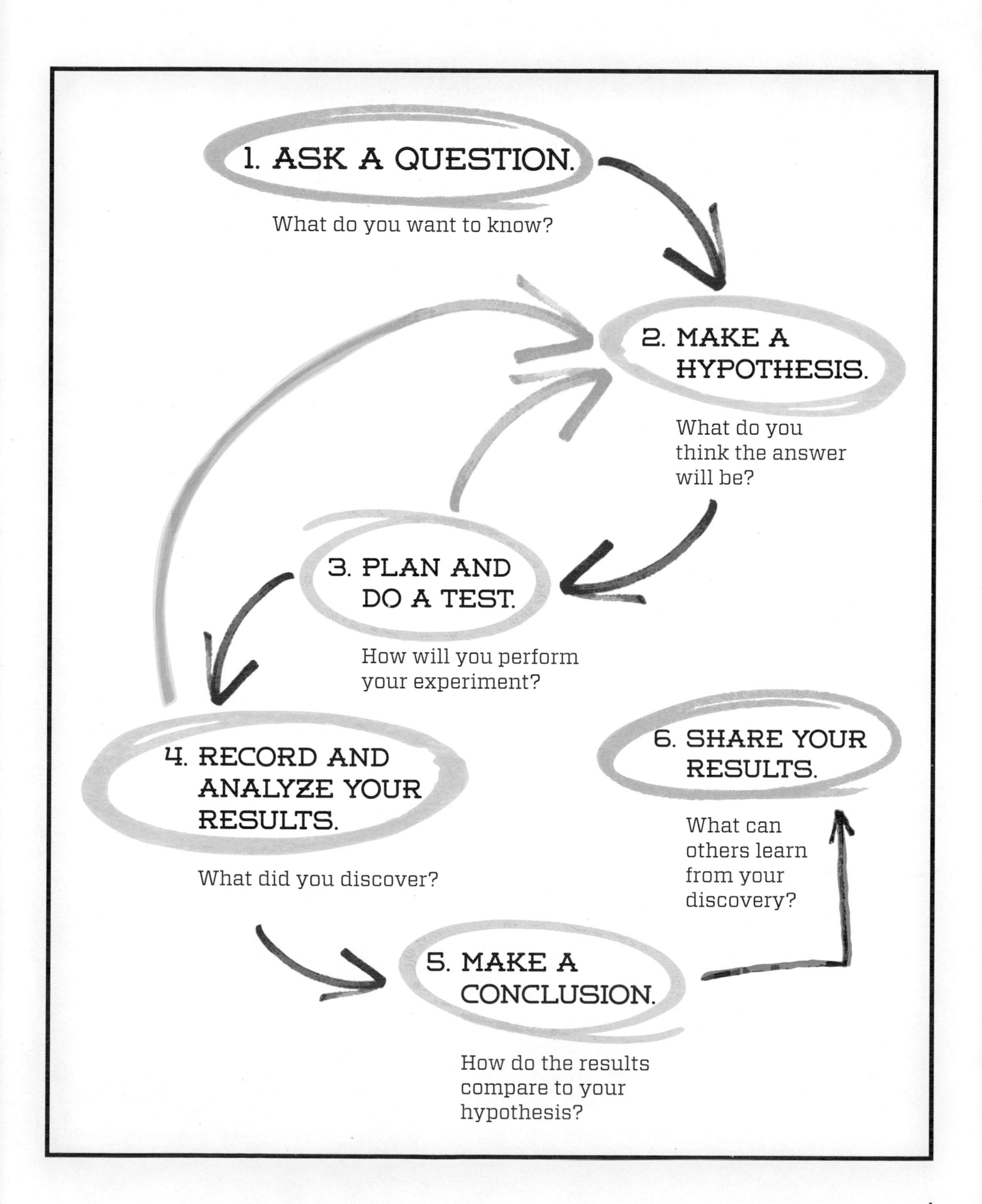
1. ASK A QUESTION.
What do you want to know?
2. MAKE A HYPOTHESIS.
What do you think the answer will be?
3. PLAN AND DO A TEST.
How will you perform your experiment?
4. RECORD AND ANALYZE YOUR RESULTS.
What did you discover?
5. MAKE A CONCLUSION.
How do the results compare to your hypothesis?
6. SHARE YOUR RESULTS.
What can others learn from your discovery?

LIFE SCIENCE: CYCLES

– the study of living things and how they interact with the natural environment

PHYSICAL SCIENCE: TRANSFORMATIONS

EARTH AND SPACE SCIENCE: PREDICTABILITY

HUMAN BODY: BALANCE

NATURAL CYCLES

VOCABULARY

biogeochemical cycle
(ˌbī·ō·jē·ə·ˈke·mi·kəl ˈsī·kəl)
the cycling of chemical elements through the living and nonliving parts of an ecosystem

KEY IDEAS

- Certain chemical elements are most abundant in living things.
- Certain elements necessary for living things cycle between the biotic and abiotic components of ecosystems.
- People can disrupt biogeochemical cycles.

Look around you at the variety of living and nonliving things in God's creation. It is amazing that everything we see is composed of one or more elements in different combinations. In fact, one of the important characteristics of organisms is that their bodies can use atoms or small molecules to construct larger molecules. These larger molecules are used to build body parts or help carry on the chemical reactions necessary for life. Oxygen (O), carbon (C), hydrogen (H), and nitrogen (N) are four elements that are most abundant in living things. These four elements move to and from living and nonliving things through biogeochemical cycles. Ecosystems remain healthy and organisms thrive because of these cycles.

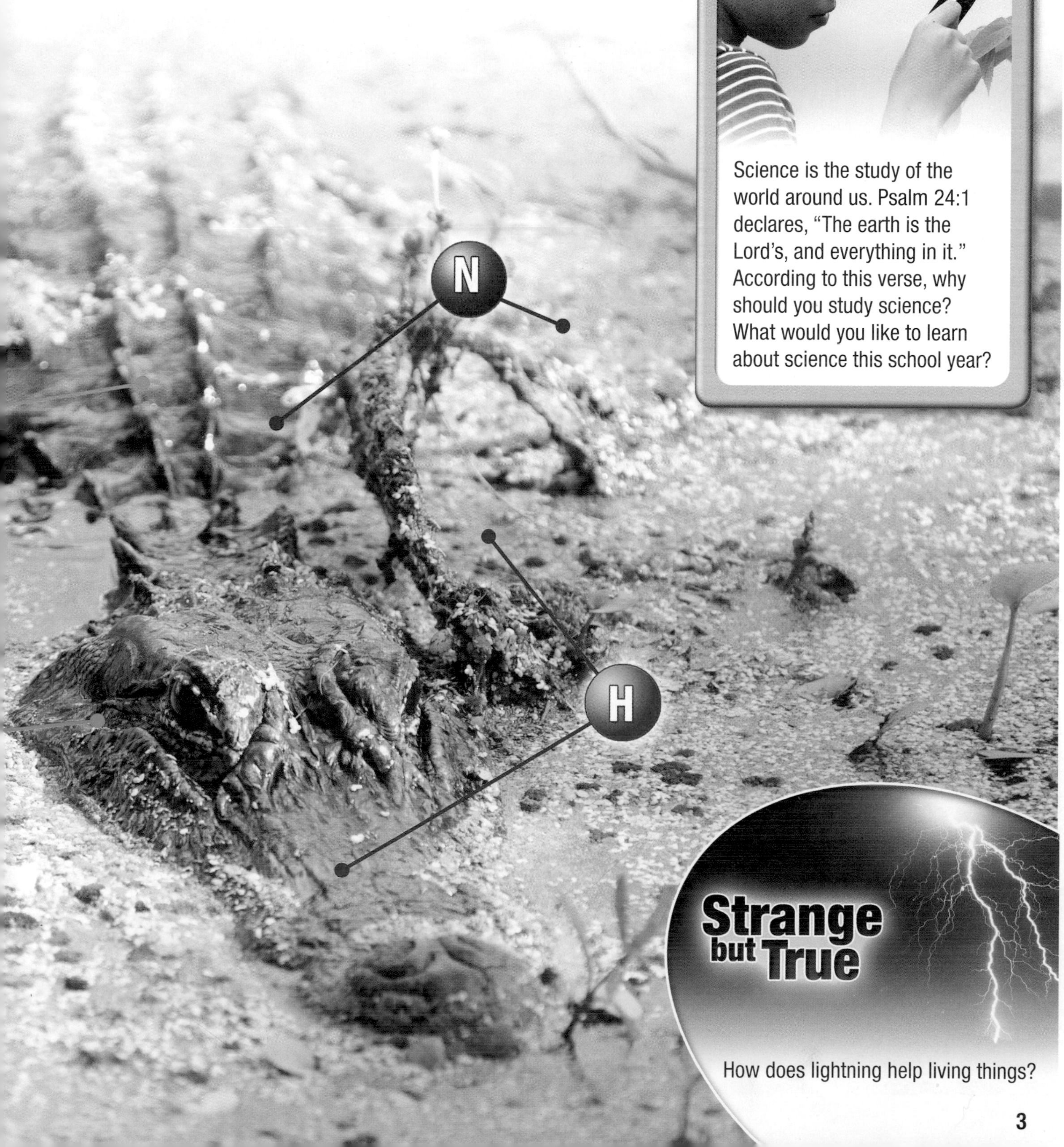

OUTLOOK

Science is the study of the world around us. Psalm 24:1 declares, “The earth is the Lord’s, and everything in it.” According to this verse, why should you study science? What would you like to learn about science this school year?

N

H

Strange but True

How does lightning help living things?

Ecosystems are composed of both biotic (living) and abiotic (nonliving) components. Biotic components depend on one another, as well as on the abiotic components of their habitat, for the elements they need to survive. Oxygen, carbon, hydrogen, and nitrogen cycle from the biotic to the abiotic factors repeatedly.

Most of the time, the elements needed for life are abundant in an ecosystem. However, in places like the desert or the open ocean, these elements may be scarce. Life is less abundant in these environmental communities. In addition, human activities sometimes interfere with the natural biogeochemical cycles, causing imbalances in ecosystems. In order to protect the planet, scientists study the causes and effects of these interferences.

Sunlight provides the energy that fuels the cycles.

Biotic components, such as producers, use sunlight and elements from abiotic components of the ecosystem to make the energy they need to grow.

The abiotic components of an ecosystem contain the elements that organisms need.

Fungi and bacteria break down dead organisms and return elements from the biotic to the abiotic components of the ecosystem.

QUICK FACT

The open ocean was once considered a chemical desert because the elements essential for life are less abundant there than they are near the shore. However, scientists have discovered that many unique organisms inhabit the ocean far from shore. In fact, even deep beneath the surface where sunlight is limited and nutrients are scarce, some spectacular creatures thrive. While these areas support fewer organisms, they are far from being a chemical desert!

The elements then pass through the biotic components as consumers eat plants or other animals.

VOCABULARY

runoff ('run·of) the precipitation that flows over the surface of soil

collection (kə·'lek·shən) the process by which precipitation is gathered into a body of water

IN THE FIELD

Snow surveyors take samples of snow by driving a sampling tube into snowpack. The snow core is weighed to estimate the water content and forecast the amount of water runoff from snowmelt. This information is important for farmers and utilities that supply water to cities.

The water cycle is mostly abiotic. Organisms, however, also contribute to this cycle. The sun's energy causes water to evaporate and enter the atmosphere as water vapor. The ocean is the greatest source of water vapor. However, water also evaporates from other bodies of water and from the leaves of plants during transpiration. Animals and humans return water to the atmosphere through their moist breath or through evaporation of their sweat or urine. Once in the atmosphere, water vapor rises and condenses to form clouds. Eventually, this water may form precipitation and fall back into the ocean or onto the land.

Some precipitation will soak into the soil and become groundwater. Water that is not absorbed into the soil moves along the surface of the ground as runoff. When this crosses the soil, it dissolves the soil's nutrients and carries them along. Runoff from the land is gathered into streams, ponds, lakes, and rivers through collection. Animals often gather to drink at these collection sites. The water may gather into rivers that flow into larger and larger rivers. Eventually, the water in the largest rivers returns to the ocean to be recycled. Some small rivers flow directly into the ocean as well.

QUICK FACT

A drainage basin is all the land area that contributes runoff to a river. The Mississippi River drainage basin is the third largest in the world. The water from the Mississippi River returns to the ocean through the Gulf of Mexico.

Will a mountain or a valley have more fertile soil? Why?

Wherever a collection of water exists, an abundance of organisms will surely follow.

This runoff is from snow melting in the mountains. It is ice cold!

CARBON AND OXYGEN

VOCABULARY

photosynthesis (ˌfō·tō·ˈsin·thə·səs) the process that allows green plants to make sugars from sunlight, water, and carbon dioxide and that releases oxygen into the atmosphere

cellular respiration (ˈsel·yə·lər res·pə·ˈrā·shən) the process in cells by which oxygen and glucose are used to produce energy and carbon dioxide

fossil fuel (ˈfä·səl ˈfyo͞ol) a source of energy formed from dead plants and animals

QUICK FACT

Plants are not the only producers. Algae and bacteria are the major producers in freshwater and marine environments. *Volvox* colonies are about the size of a grain of sand. This is *Volvox*, a green alga.

Every organism is made of molecules that contain carbon. Most organisms also use oxygen. Carbon and oxygen repeatedly cycle from the atmosphere, through a chain of organisms, and back to the atmosphere. The processes of photosynthesis and respiration are part of this cycle.

The body of an organism uses oxygen to break down sugar, protein, and fat molecules to release the energy stored in them. These molecules contain carbon. The organism uses the energy to live and grow and releases carbon dioxide into the atmosphere. This process is called cellular respiration.

When a herbivore consumes plants, it acquires the carbon that it needs. Carbon continues up the food chain when a carnivore consumes the herbivore. The transfer of carbon occurs repeatedly as one organism then consumes another in the food chain.

Oil, coal, and natural gas are commonly called fossil fuels. They are the remains of organisms that have been deposited within the earth's crust. When people burn fossil fuels to generate electricity, heat homes, or fuel cars, carbon is released as carbon dioxide into the atmosphere.

When organisms die, bacteria and fungi decompose the bodies. Decomposition releases carbon dioxide into the atmosphere.

Producers absorb carbon dioxide from the atmosphere. During photosynthesis, they use carbon dioxide, water, and sunlight to produce glucose. Producers use the energy stored in the sugar to grow, develop, and reproduce. They release oxygen into the atmosphere. All carbon enters the biotic part of the ecosystem through a producer.

NITROGEN

VOCABULARY

nitrogen fixation (ˈnī·trə·jən fik·ˈsā·shən) the conversion of nitrogen in the atmosphere to a form that can be used by plants

legume (ˈle·gyōōm) a plant that hosts nitrogen-fixing bacteria in nodules on its roots

nodule (ˈnä·jōōl) a swelling on a plant root that contains nitrogen-fixing bacteria

denitrification (dē·ˌnī·trə·fə·ˈkā·shən) the process that releases nitrogen from soil back into the atmosphere

Nitrogen is another essential element for organisms. The chlorophyll that makes plants green, the hemoglobin that makes blood red, and the protein in muscle tissues all contain nitrogen. Although about 78% of the atmosphere is nitrogen, it is not in a form that most organisms can use. Nitrogen must first be combined with other elements to form compounds that can be utilized. This process is called nitrogen fixation. It is performed by certain kinds of bacteria, many of which live in plants.

Legumes are certain kinds of plants that have small bumps, called nodules, on their roots. Special bacteria that live in the nodules have an amazing ability. They take nitrogen that has come into the soil from the atmosphere and change it into a form that the plants can use for growth and reproduction.

When a herbivore eats a legume, it obtains the nitrogen it needs to live and grow.

Strange but True

Lightning is so powerful that when it flashes through the air it can change nitrogen gas into a form that organisms can use. This is called *atmospheric nitrogen fixation.* The new form of nitrogen dissolves in rainwater. This means that rainwater from thunderstorms can be a mild plant fertilizer!

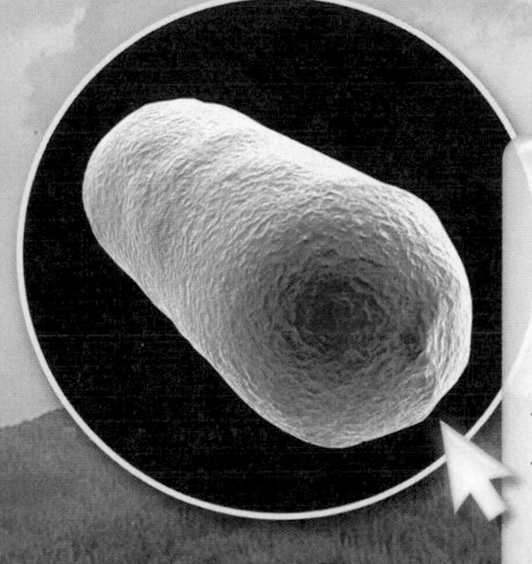

Some bacteria take the nitrogen in the soil and return it to the atmosphere through a process called denitrification. When an ecosystem is in balance, the amount of nitrogen that is added to the biotic part of the ecosystem by nitrogen fixation is almost equal to the amount of nitrogen that leaves through denitrification.

QUICK FACT

Bacteria are not the only things that fix nitrogen. Chemists apply an industrial process to fix atmospheric nitrogen into a useful form. They produce fertilizers in this way. Fertilizer makes crops grow faster and produces more food for people. However, when fertilizers are washed away by runoff, they can become water pollutants.

When an animal or plant dies, bacteria and fungi decompose the body, returning the nitrogen to the soil. Other plants then use this nitrogen, and the cycle in the biotic part of the ecosystem starts again.

When a carnivore eats a herbivore, the nitrogen is passed up the food chain. These animals return nitrogen to the soil through their nitrogen-containing urine and feces.

POLLUTION

VOCABULARY

photochemical smog (ˌfō·tō·ˈke·mi·kəl ˈsmäg) the brown smog produced when air pollutants react with sunlight

emission (ē·ˈmi·shən) a pollutant released into the atmosphere

acid precipitation (ˈa·səd pri·ˌsi·pə·ˈtā·shən) the precipitation that is more acidic than normal rainwater

God designed the carbon and oxygen cycle and the nitrogen cycle to be balanced in order to allow life to thrive on Earth. However, humans sometimes disrupt these cycles. Carbon and nitrogen become pollutants when too much of them are introduced into an ecosystem. When fossil fuels are burned in electric power plants, cars, jets, and even lawn mowers, polluting gases enter the atmosphere. These gases are hydrogen and carbon compounds and nitrogen and oxygen compounds.

When sunlight interacts with these atmospheric pollutants, it forms chemicals that can make eyes, noses, and throats burn. These chemicals make up photochemical smog, and nitrogen dioxide is one of its major components. Smog is harmful to the environment and to human health. Some countries like Mexico and Greece use an unusual approach to reduce the pollution that causes smog. They limit the number of vehicles that enter major cities. Other nations control the level of emissions produced by cars and factories.

If you have ever flown in an airplane, you may have seen a brown haze hanging over some cities. The burning of fossil fuels causes most photochemical smog. Cars and factories are more abundant in large cities, which can lead to poor air quality.

When some nitrogen pollutants combine with water vapor in the atmosphere, they change the water droplets from being mildly acidic to strongly acidic. Air currents blow these acidic droplets far from where they were produced. These droplets condense and fall to the earth as acid precipitation. Acid precipitation may collect in streams, rivers, and lakes. In fact, some lakes have become so acidic that fish can no longer live there.

Acid precipitation soaks into the soil and can destroy the ability of trees to absorb the nutrients they need. This stresses the trees, making it difficult for them to survive attacks by insects or diseases. Some forests have been severely damaged by acid precipitation. However, people are finding ways to lessen the pollution that causes acid precipitation. They use public transportation or carpool to reduce the number of cars on the road. Others walk or ride bikes to school or work. Some companies want to create less air pollution and so are researching new methods of making their products.

Why would scientists at Hubbard Brook Experimental Forest treat the soil in one watershed with a calcium-containing mineral?

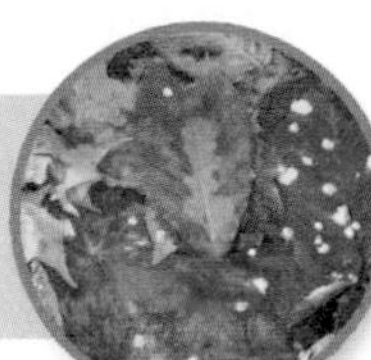

LINKS

History Link
Industrial smog is another kind of smog. Research the clean-up effort that Beijing, China implemented to improve the air quality for the 2008 Beijing Olympics.

Language Link
Research and write a paragraph about the advantages and disadvantages of using fertilizer. Include a description of how fertilizer can become a pollutant and what other methods could be used to grow crops.

Art Link
Collect objects from nature and mount them on poster board to create a montage illustrating one of the biogeochemical cycles.

Acid precipitation does not taste or smell different than normal precipitation. However, its effects can be seen. Over time, these statues have been eaten away by acid precipitation.

Some scientists are researching new kinds of fuel that will reduce air pollution.

CHAPTER 2

LIFE CYCLES

All living things go through stages that occur repeatedly in the same order. Scientists usually refer to a whole species' life cycle as one that begins with an egg or spore and continues until a mature adult reproduces offspring. If an organism does mature to adulthood, it often changes in size, shape, or function during its development. However, on an individual level, a single organism might die before completing all the stages of a particular species' life cycle. One complete life cycle is called a generation.

God designed all organisms with a set of instructions for completing their life cycles. This information is the genetic material contained in every cell of an organism. Genetic material is like a recipe book that gives detailed directions about growing and developing. When an organism reproduces, this genetic material is passed on to its offspring. The next generation then possesses instructions for its own life cycle.

VOCABULARY

life cycle (ˈlīf ˈsī·kəl) the series of stages that an organism passes through from egg or spore to reproducing adult

generation (je·nə·ˈrā·shən) one complete life cycle

genetic material (jə·ˈne·tik mə·ˈtîr·ē·əl) the set of instructions within a cell that controls an organism's characteristics and life processes

KEY IDEAS

- All living organisms exhibit life cycles.
- An organism's genetic material directs how the organism grows and develops throughout its life cycle.
- Different kinds of organisms have different life cycles.
- Sometimes human activities interfere with the life cycle of other organisms.

Tadpoles eventually lose their tails, develop lungs, and become adult frogs. Adult frogs produce eggs that develop into the next generation.

Life Cycle of a Tree Frog

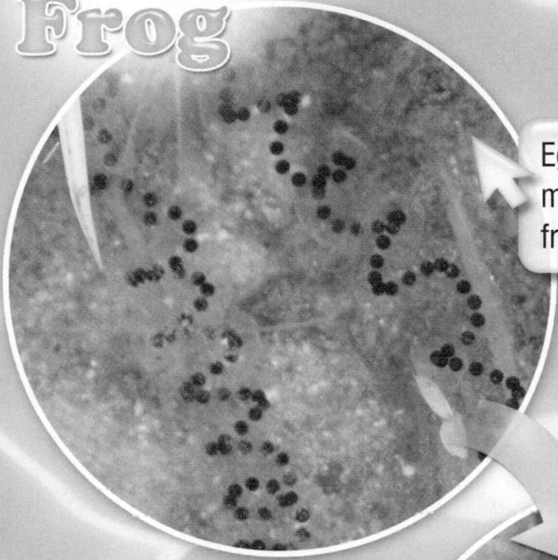

Eggs contain the genetic material needed for the frog to grow and develop.

Some young organisms, like tadpoles, do not look much like their parents. They use gills to breathe underwater.

Tadpoles start to grow legs and develop in other ways.

IN THE FIELD

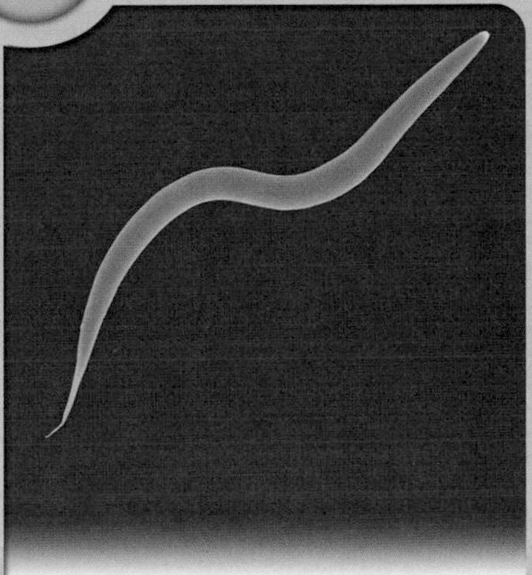

This is a roundworm called *C. elegans*. It is transparent and tiny—only about 1 mm (0.04 in.) long. Because the worm completes its life cycle in approximately three weeks, it is easy for scientists to study. Biologists have used this worm to try to better understand how genetic material affects reproduction, growth, and aging. The *C. elegans* roundworm has even gone into outer space with the French Space Agency to be observed and analyzed.

Strange but True

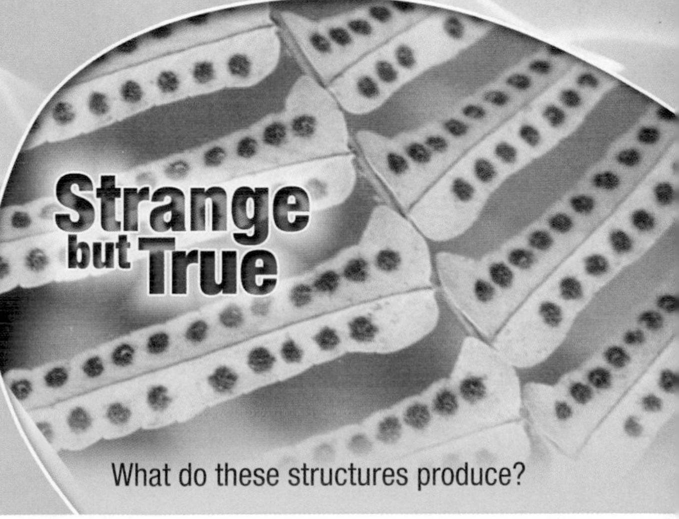

What do these structures produce?

BACTERIA

OUTLOOK

Modern science uses technology to study bacteria so people can be educated about how to avoid harmful bacteria. God's laws in the Old Testament instructed the Israelites how to prepare foods, which foods to avoid, and what health practices would protect them from harmful bacteria.

One of the simplest life cycles belongs to some of the smallest organisms on Earth—bacteria. Humans are made up of many cells, but bacteria are *unicellular*, which means *made of only one cell*.

Bacteria are so small that they can only be seen with a microscope. Although they are tiny, they live everywhere! Bacteria live on desks, in snow, in the atmosphere 64 kilometers (40 miles) above the earth, and in hot thermal vents. They also thrive inside, and on, other organisms. Bacteria are in your mouth and the pores of your skin, as well as in stomachs of bison, in nodules on the roots of peanuts, and on leaves that are decomposing. Bacteria can live below the surface soil of the earth and beneath the ocean floor. Even though some bacteria are harmful, most are not. Researchers are currently listing all the different species that they find.

Bacteria live in interesting and sometimes unexpected places.

How many bacteria are there on and below the surface of the earth?

Under ideal conditions, one well-known species of bacteria (*E. coli*) completes its life cycle in only 20 minutes by simply dividing in two. First, a copy of the bacterium's genetic material becomes visible. Then the cell starts to elongate, separating the two copies of the genetic material. A new cell wall forms down the middle, dividing the cell into two separate cells. These next generation cells are smaller than the cell they came from, but they are almost identical to the first, or mother, cell. Their genetic material controls this process of cell growth and division as the cycle continues.

Bacteria can sometimes multiply very quickly. With the right temperature and enough food, they can thrive, reproducing rapidly. Fortunately, conditions are not always perfectly suited for their growth.

QUICK FACT

Your hands harbor bacteria that can reproduce and make you sick. Hands contaminated with bacteria can spread a variety of diseases. Studies have shown that most of the bacteria on your hands can be killed by simply washing with soap and warm water for about 15 seconds. It is important to clean under your nails as well as between your fingers.

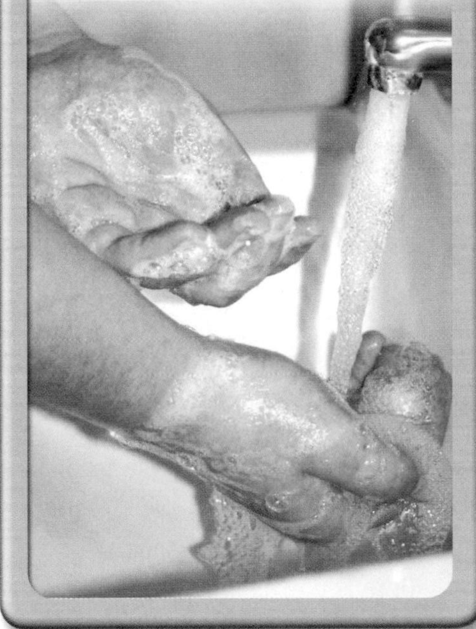

How does penicillin fight bacteria?

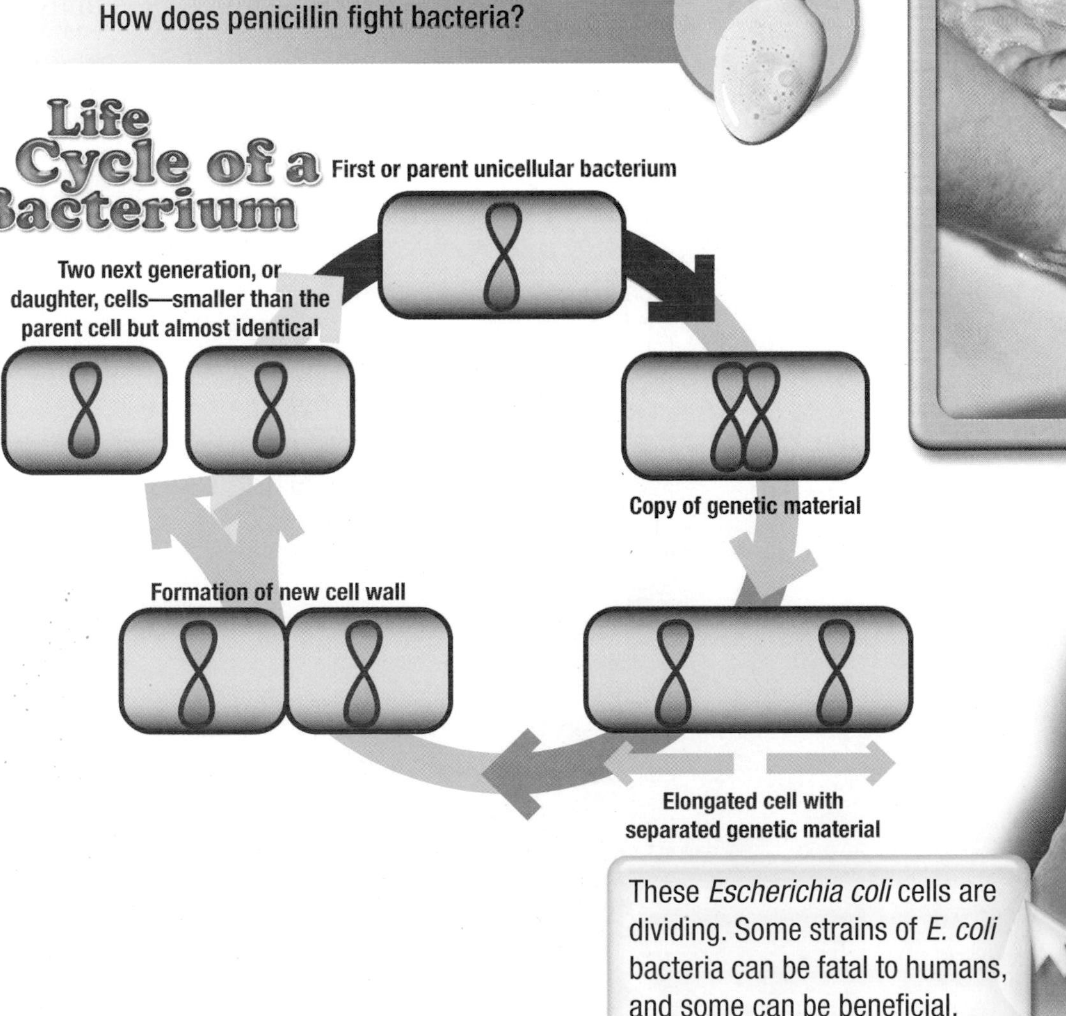

These *Escherichia coli* cells are dividing. Some strains of *E. coli* bacteria can be fatal to humans, and some can be beneficial.

FUNGI

VOCABULARY

fungus (ˈfung·gəs) a decomposer that usually reproduces through spores

mushroom (ˈmush·rōōm) a fleshy, spore-producing growth of certain fungi

gills (ˈgilz) the structures on the underside of a mushroom

IN THE FIELD

Mycologists are scientists who study fungi. They understand which mushrooms are edible and which are not. Although people often refer to poisonous mushrooms as toadstools, there is no scientific distinction between mushrooms and toadstools. Mycologists warn people to never eat mushrooms they find on the ground—and for good reason. For instance, this fungus is called *death cap*. Eating as little as one half of the cap can cause death.

A fungus is neither a plant nor an animal. It is an organism that fits into a completely separate category. There are over 50,000 types of fungi, including molds, mildews, rusts, and yeasts. Some, such as the yeast used to make bread, are unicellular. But most, like the mold that grows on bread, are multicellular.

A fungus usually reproduces through spores and gets its energy from either living or once-living things. It is a decomposer that releases enzymes into organic matter. These enzymes break down the available nutrients found in the organic matter. The fungus then absorbs some nutrients to power its growth.

Fungi grow almost everywhere, but they thrive in warm, moist areas. They are an important part of the ecosystem since they decompose dead organisms, returning nutrients to the soil. They are used to create food, cheese, and antibiotics. Although most fungi are not harmful, some can cause disease, spoil food, or destroy plants.

What can make feet itch?

The holes in bread are formed by baker's yeast. The yeast cells interact with the sugar and flour in the dough to make carbon dioxide. The bubbles from the carbon dioxide gas cause the dough to rise.

Fungi are decomposers that return nutrients to the soil and atmosphere as they feed on living and once-living things.

One kind of fungus that grows in the soil produces a familiar structure, the mushroom. Its main parts are the stem, cap, ring, and gills. The gills are located on the underside of the mushroom. They produce microscopic cells called spores that contain the fungus's genetic material. These spores are light and dry, so they blow away easily.

If the spores land on soil that has the right conditions, the spores will germinate and grow into a mat of long, threadlike cells. The threads usually grow hidden beneath the soil. Sometimes the threads of two mats meet. When this happens, the cells may join, creating a new mat with cells that carry genetic material from both parents. The threads in this new mat may create a mushroom. If so, the mushroom will break above the surface of the soil, making new spores that blow away to create the next generation of fungus.

QUICK FACT

Sometimes fungi form mushrooms that are arranged in a circle or half circle. This happens because the mushrooms are only growing on the very edge of an underground mat of threadlike cells. These mats of cells grow in all directions at the same time. However, the mushrooms may appear only on the edges in circular or semi-circular formations called *fairy rings*.

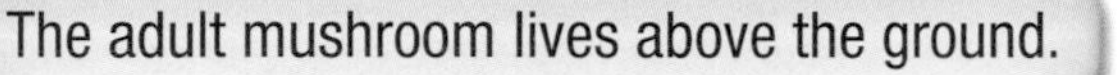

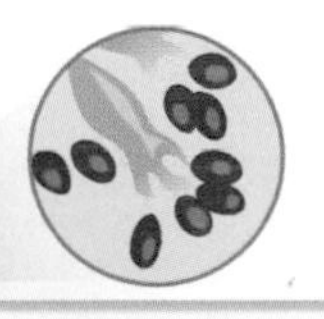

The spores of the *Lycoperdon* mushroom, commonly called *a puffball*, disperse into the air with great energy. Falling rain or an animal passing by can cause the mushroom to puff out its spores.

VOCABULARY

cotyledon (kä·tə·ˈlē·dən) the nutrient-rich structure in a seed

embryo (ˈem·brē·ō) an organism produced from a spore or fertilized egg, in the early stages of development

seedling (ˈsēd·ling) a young plant

fertilization (ˌfər·tə·lə·ˈzā·shən) the uniting of the genetic material of two cells

fruit (ˈfro͞ot) a plant structure that surrounds the seed(s) of a flowering plant

Plants have different types of complex life cycles, but a bean plant is an example of a fairly common cycle. A bean seed has three parts. The seed coat covers the seed and helps it survive until conditions are just right for growth. The cotyledons are packed with nutrient-rich molecules to fuel the early plant's growth. The embryo eventually becomes the new plant when a seed germinates.

When a bean seed is planted where conditions are ideal, it often germinates. The first structure to appear is the root. Later, the seed coat splits. Then a stem develops and grows up toward the sunlight. This new stem pulls the cotyledons out of the ground with it. The little seedling develops its first leaves and begins to make its own food through photosynthesis. It eventually develops into an adult plant with bigger leaves and flowers.

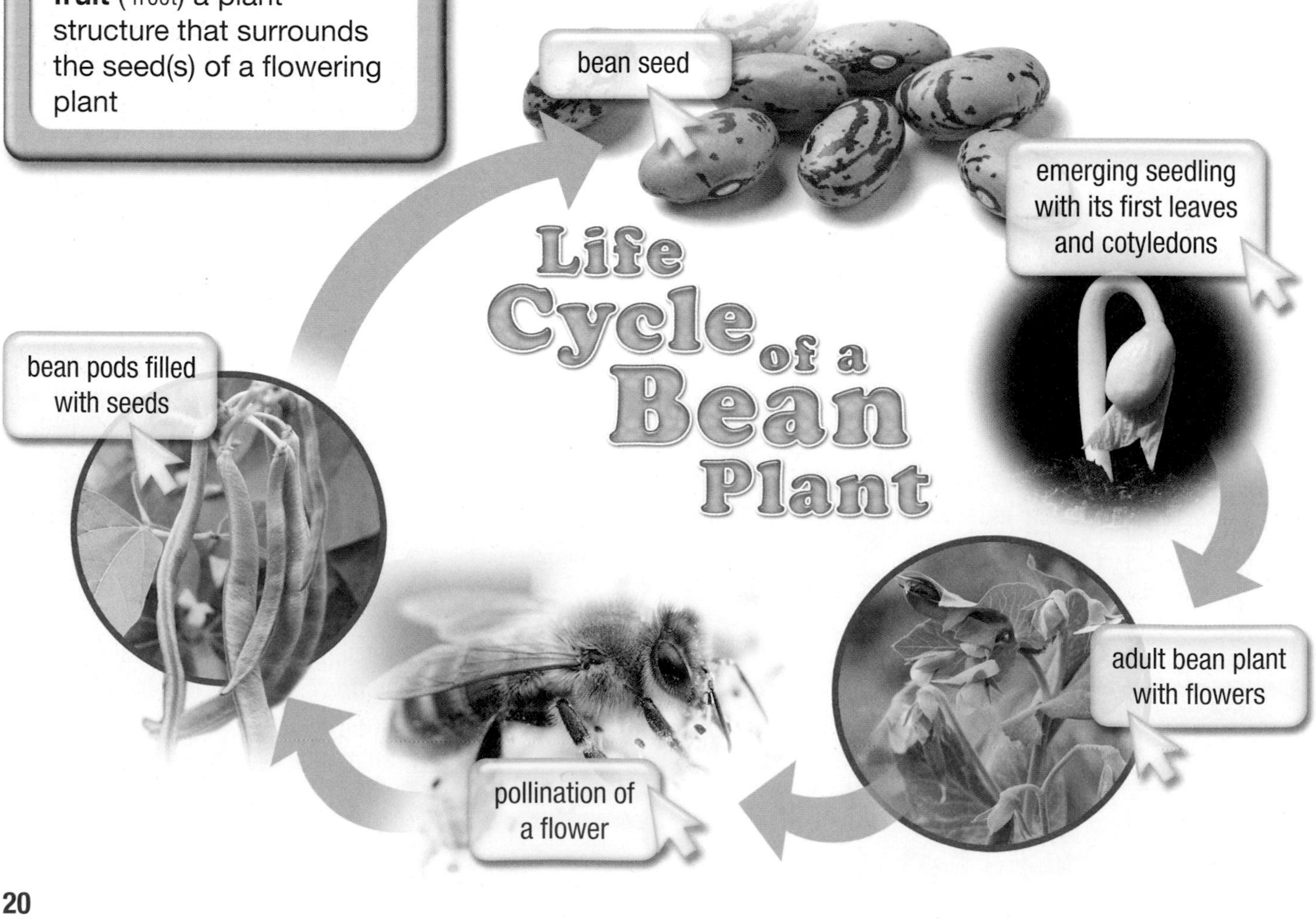

Flowers are the reproductive structures of the plant. The top part of the flower's stamen produces pollen. Each pollen grain contains half of the plant's genetic material. Pollen grains are carried from one plant to another by pollinators. The pollen sticks to the top portion of the flower's pistil. Then the pollen grows down into the base of the flower where there are tiny structures containing eggs. Each egg carries the other half of the plant's genetic material. When the pollen grain merges with the egg, their genetic material mixes and fertilization occurs. The fertilized egg becomes a seed that is enclosed within the bean pod. Although the bean pod is not sweet like an orange, it is the fruit of the bean plant that develops from the bottom part of the pistil. If a seed from the pod is planted and germinates, the embryo inside the seed becomes the next generation plant.

How can scientists find out about which plants lived thousands of years ago?

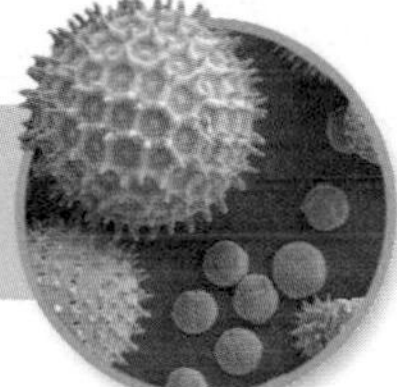

QUICK FACT

Saguaro cactus flowers contain sweet nectar, which appeals to certain bats. The bat pokes its muzzle deep into the flower. As it drinks, the bat's head rubs against the stamen of the cactus flower and becomes covered with pollen. When the bat travels to another cactus flower, it delivers the pollen to the stigma of that flower. This pollination process allows the flower to produce seeds, completing the cactus' life cycle.

Lima Bean Flower

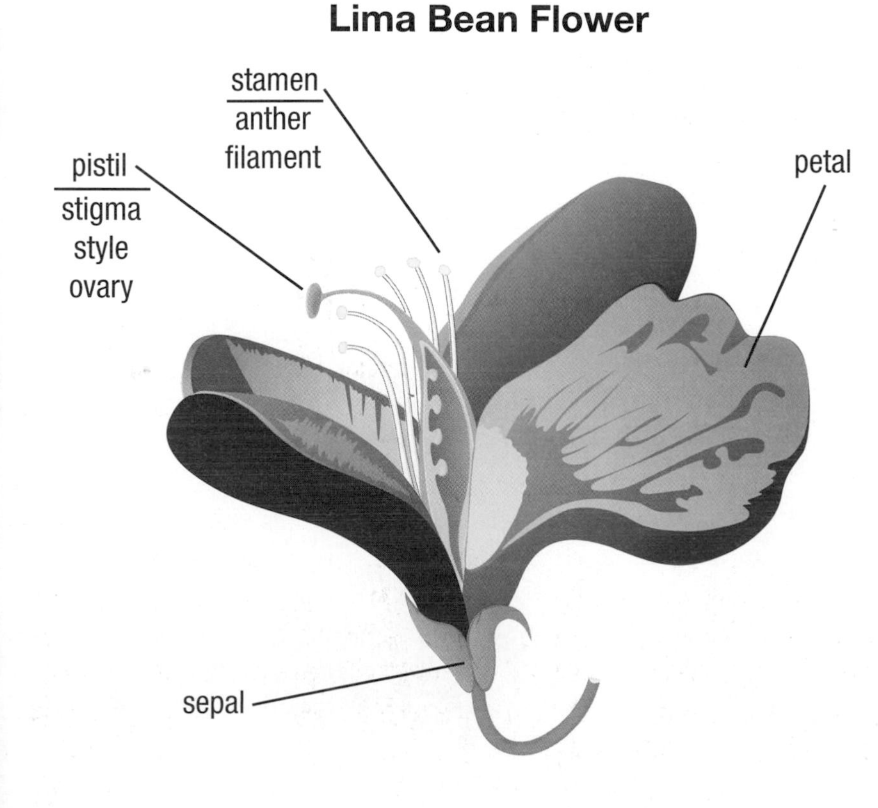

Not all plants produce seeds. Ferns produce single-celled spores in round brown structures, found on the underside of the fern frond. The spores are light and dry and blow away. If the spores land on soil, under the right conditions, the spores will grow and produce a new fern generation.

SALMON

VOCABULARY

redd ('red) a salmon nest

alevin ('a·lə·vən) a very young salmon hatchling

parr ('pär) a young, growing salmon

smolt ('smōlt) an immature salmon that migrates to the ocean

Salmon are vertebrates whose life cycles are unique, as well as complex. They begin their lives in freshwater streams, travel down rivers to the ocean, and then usually return to the same freshwater streams to reproduce. Salmon live in the Pacific Ocean, the Atlantic Ocean, the Bering Sea, the North American Great Lakes, numerous streams and rivers, and even in some landlocked lakes.

Adult female salmon lay their eggs in the late summer or early fall in a nest called a redd. Baby salmon, called alevins or sac fry, hatch in early spring. Alevins use their small yolk sacs containing nutrient-rich molecules to sustain growth. When the yolk sac is gone, the salmon eventually enter the parr stage. The salmon parr live in the stream and river, eating mostly insects, until they are one to three years old. At this stage they are called smolts. Smolts travel down the river, spend several months in water that is a mixture of salt and freshwater, and then swim into the ocean and mature into adults.

eggs

alevin

adult

Life Cycle of a Salmon

smolt

parr

The adult salmon feed on other fish, squid, and crustaceans to store nutrients for their return trip. Adult salmon try to find the mouths of the rivers that they swam down as smolts. Some scientists think that the earth's magnetic field acts as a compass to direct the fish there. The salmon must then swim up the river to find the streams where they hatched, probably using their sense of smell. Some Alaskan salmon swim over 1,610 km (1,000 mi) and do not eat for the entire trip!

Once at the stream, the female creates several redds in the gravel by lying on her side and thrashing her tail. There she lays her eggs, each of which contains one half of the fish's genetic material. A male salmon then moves over the eggs, depositing cells that have the other half of the salmon genetic material, fertilizing the eggs, and completing the life cycle.

IN THE FIELD

Some species of adult salmon die shortly after laying and fertilizing their eggs, but their bodies give an amazing gift to the land. Some salmon are eaten by eagles and bears. However, the bodies of others decompose in the riverbeds. The stored nutrients in the salmon are deposited in the soil surrounding the rivers. Plants use these nutrients to grow. Then land animals eat the plants.

What is a fish ladder?

Salmon can grow quite large from all the nutrients they consume from ocean life. The largest Chinook salmon ever recorded was caught in 1949 and weighed about 57 kg (126 lb)! It was 147 cm (58 in.) long.

Dams, farm irrigation, logging, pollution, and overfishing have contributed to the endangerment of some salmon species.

MALARIA

QUICK FACT

Malaria is often present in countries that surround the equator.

Scientists use their understanding of life cycles to help people. Malaria is a disease that makes hundreds of millions of people sick each year. Over half a million people die, most being children. Malaria means "bad air," for many once thought that this disease came from breathing air around swamps. Actually, a small parasite transferred by mosquitoes causes malaria.

A mosquito uses its proboscis, a pointy structure on its head, to puncture the skin of a person or animal. It then injects its saliva. An infected mosquito may pass the malaria parasite into the person's bloodstream. This parasite infects the person's liver and red blood cells. Soon he or she experiences a high fever that comes and goes repeatedly. The next mosquito that bites the infected person withdraws a drop of blood containing the malaria parasite. The parasite multiplies in the gut of this mosquito and then moves to its mouth. When the mosquito bites again, the parasites may be injected into another person.

Plasmodium falciparum, a one-celled organism, is the parasite responsible for about 80% of malarial infections in humans. This microscopic image shows human blood cells with *Plasmodium falciparum* at various growth stages.

Do all mosquitoes carry malaria?

This female *Anopheles* [ə·ˈnä·fə·lēz] mosquito is ingesting a blood meal. It may infect the next person it bites with malaria.

One way to prevent the spread of malaria is to control the mosquito population. Mosquitoes must lay their eggs in standing water. Emptying filled buckets, covering water jars, and getting rid of old tires that collect rainwater help lower the number of mosquitoes. Sometimes people spray insecticides or sleep under insecticide-sprayed mosquito nets. Unfortunately, some insecticides have been used so much that they no longer get rid of the mosquitoes that carry this disease.

There are medicines that help infected people. However, malaria often affects those who can least afford to buy the necessary medicines. To complicate matters the malaria parasite has recently become resistant to many of these drugs. The good news is that many people are trying to develop other ways to help combat this disease. Soon scientists hope to create a vaccine that would make humans immune to malaria and save thousands of lives every year.

What tree are these leaves from?

LINKS

Language Arts Link
Research the fungus *Armillaria bulbosa*. Write a paragraph describing where it grows, how big it grows, what the mushroom looks like, and how researchers found it.

Art Link
Create a poster that illustrates at least five ways students can avoid spreading bacteria at school.

Social Studies Link
Present an oral report on state, federal, or international laws that protect salmon; or, on community groups that are trying to create salmon spawning areas.

Geography Link
Research salt lakes around the world and what lives in them. Use an outline of a world map to identify and label the lakes.

About 85% of all malaria deaths occur in Africa. Sadly, many of those who die are children. Countries near the equator have temperatures above 20°C (68°F), which is the perfect environment for mosquitoes.

CHAPTER 3

CELLS

KEY IDEAS

- All living things are made of cells.
- Cells are the basic units of structure and function of all living things.
- All cells come from other cells.
- Cells contain smaller structures called *organelles*.
- Cells divide and specialize.

Imagine a large building, such as a skyscraper or castle. What is it made of? What are its smallest parts? Often, individual blocks fit together to make larger formations. In tall buildings, cinder blocks and mortar or steel and drywall are used to make working units. However, the smallest unit of structure and function can be something as basic as a brick.

In Life Science the concept of the cell is quite similar to the bricks in a building. Both are the basic parts of the larger system they form. Cells, therefore, are considered the basic unit of structure and function in all living organisms. By studying cells and what is inside them, scientists can learn more about the organisms they make up.

Masons lay stones one by one to create amazing formations like these walls of Jerusalem. In a similar way, cells are the basic units that make up all living organisms. Without cells, organisms would not have any structure or function. If its cells do not work properly, neither does the organism.

IN THE FIELD

A microbiologist is a scientist who studies microbes and how they affect the world. Microbes include bacteria, viruses, and fungi. Microbiologists test the food people eat, develop medicines, help control the spread of disease, and even ensure that laundry detergents will not irritate skin.

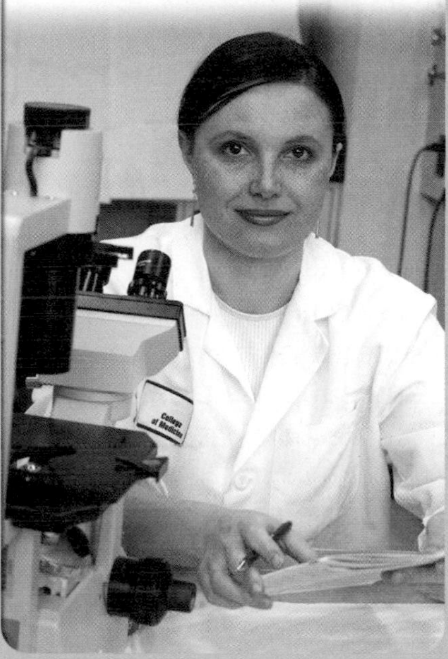

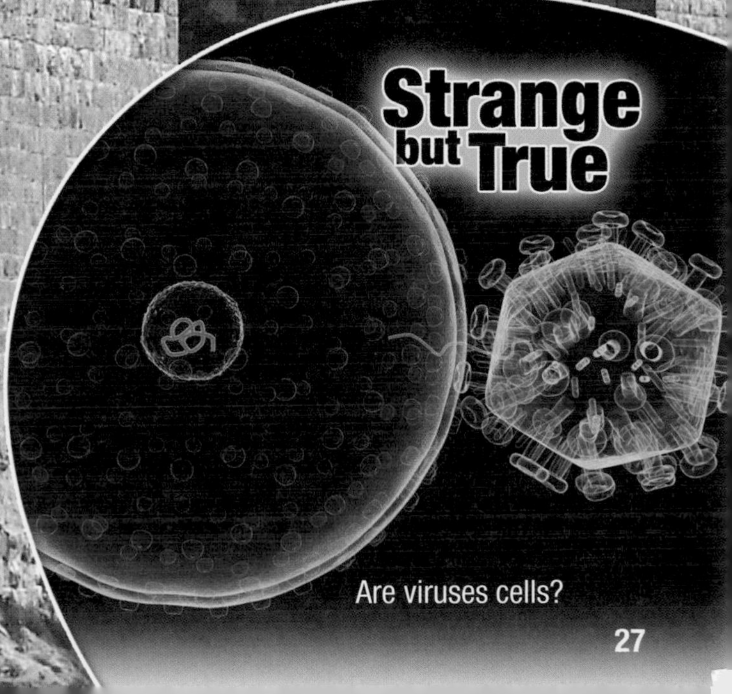

Strange but True

Are viruses cells?

CELL STRUCTURE

VOCABULARY

organelle (ôr·gə·ˈnel) one of several tiny structures within a cell

cell membrane (ˈsel ˈmem·brān) a flexible structure that protects and controls what goes in and out of the cell

cytoplasm (ˈsī·tə·pla·zəm) the jelly-like substance found inside the cell that contains organelles

cell wall (ˈsel ˈwol) the rigid outer layer that protects and supports plant cells

It was not always known that all living things are made of cells. The invention of the microscope made it possible for people to observe and study cells. English scientist Robert Hooke was one of the first people to view cells. In 1663 he built a compound microscope and used it to observe cork. Soon, many people began studying cells. From the discoveries of three German scientists named Matthias Schleiden, Theodor Schwann, and Rudolf Virchow, the cell theory was developed. The cell theory states that all living things are composed of cells, cells are the basic unit of all organisms, and all cells come from other cells.

Cells are made of smaller parts called organelles. Each organelle has a specific job that helps keep the cell alive. Plant and animal cells have a control center, or nucleus, which contains the genetic instructions for the cell. This determines the kind of cell, what it will look like, and the function it will perform.

Even though cells are extremely small, organelles are even smaller. The green image shows rough endoplasmic reticulum and the purple image displays mitochondria and ribosomes—all organelles within cells.

All cells have a cell membrane. The cell membrane is a flexible structure that protects the cell. It forms the outer boundary of an animal cell. The cell membrane selects which materials will pass in or out of the cell. All cells also contain cytoplasm. Cytoplasm is the jelly-like substance found inside the cell membrane. It contains the cell's organelles.

Plant cells also have cell walls, which animal cells do not have. The cell wall is a rigid layer of nonliving material that surrounds the cell membrane, forming the plant cell's outer boundary. Cell walls are what make fruits and vegetables crunchy. Although they are tough, selected materials like water and oxygen can pass through them easily. Plant cells are usually box-shaped, which allows the cells to stack up on each other. This, along with the rigid cell wall, provides the extra support and protection that plant cells need.

How does a cell membrane work like this door?

QUICK FACT

In the late 1830s Schleiden and Schwann made important discoveries about cells and living things. They noticed similarities between plant and animal cells. This caused scientific study to shift focus to life at the cellular level. In 1855 Rudolf Virchow proposed that cells came from pre-existing cells. The discoveries of these three scientists make up the cell theory.

Plant cells in a *Strelitzia* leaf are box-shaped.

Animal cells in kidney tissue have an irregular shape.

GENETICS

VOCABULARY

DNA (dē·en·'ā) [deoxyribonucleic acid] the molecule that contains the genetic material of a cell

chromosome ('krō·mə·sōm) a structure located in the nucleus that contains DNA

gene ('jēn) a segment of DNA located on the chromosome that, among other things, controls specific traits

trait ('trāt) a characteristic that can be passed on to an organism's offspring

DNA is the genetic material found in every cell. It is a complex molecule that carries all the information about the organism. DNA not only determines the type of each cell, but also directs all of the cell's functions.

In organisms that do not have a nucleus, such as bacteria, the DNA is located in the cytoplasm. If you look at an animal or plant cell under a microscope, you will see a dark spot. This dark spot is the nucleus, which is where the DNA is found. DNA forms into structures called chromosomes. Chromosomes often appear in the shape of an X. Each organism has its own set and number of chromosomes. Humans have 46 chromosomes, while potatoes have 48 and dogs have 78. As you can see, the number of the chromosomes does not determine the size or complexity of the organism.

Do the cells of this organism have a nucleus?

DNA is shaped like a twisted ladder. It is made of many pieces that fit together in different patterns. Each pattern is a code that gives the cell instructions for each unique trait.

Chromosomes often form into the shape of an X. Every chromosome is made of DNA, which carries the traits for each organism.

Chromosomes have many genes located on them, which are joined together like beads on a string. Each gene consists of a segment of the DNA strand. All genes carry a code and are the factors that control an organism's traits. These traits make an organism look the way it does. Although human body cells have only 46 chromosomes, they have approximately 20,000–25,000 genes! Each one of these genes is responsible for the traits you have.

You probably have more traits than you realize, from the color of your hair and eyes to the way your food is digested. You probably do not even know that some of your traits are considered traits. Smile. Do you have dimples? Does your thumb bend toward you or stay straight? Do you see all colors or have trouble recognizing some of them? These are just a few examples of the many traits controlled by genes that you have received from your parents.

OUTLOOK

Each person is a unique individual. Whether born to biological parents or adopted, God created each person in a unique way! Modern technology is showing that identical twins have slight variations in DNA, contradicting the belief that identical twins' DNA is 100% identical. God has a plan for your life that is just as unique as your individual DNA—and you are the only one who is perfectly suited to pursue His plan for your life!

Having freckles is a trait controlled by genes. For those with the freckle trait, sunlight triggers freckles to appear. Usually, these brownish spots appear after the age of five and fade somewhat in adulthood.

Most biologists do not consider viruses to be true living organisms. The only life function they can perform is reproduction. Unlike organisms, they must invade living cells to reproduce. Therefore, they are not considered cells.

THE CELL CYCLE

VOCABULARY

cell cycle ('sel 'sī·kəl) a series of events occurring during the life of a cell

interphase ('in·tər·fāz) the stage of cell growth occurring at the beginning of the cell cycle

mitosis (mī·'tō·səs) the stage of the cell cycle in which the cell's nucleus divides in two

cytokinesis (ˌsī·tō·kə·'nē·səs) the final stage of the cell cycle in which the cytoplasm divides

QUICK FACT

Different types of cells require different lengths of time to complete a cell cycle. It is estimated that adults have trillions of cells in their body at any given point in time. Special scavenger cells in humans clear away dead cells, which eventually exit the body.

Most living things do not grow larger because their cells get bigger. Instead they grow when they produce more and more cells. Interestingly, cells multiply by dividing.

The cell cycle is the series of events occurring in the life of a cell. This includes its growth, duplication of DNA, and eventual cell division. The time required to complete a cell cycle differs according to the type of cell. The common bean cell requires about 19 hours to complete a full cycle whereas yeast cells only require about 90 minutes. The cell cycle consists of three basic stages: interphase, mitosis, and cytokinesis.

Interphase is the stage in which the cell grows, develops, and matures. This occurs at the beginning of the cell cycle and takes up the most time. Here, the organelles are copied, producing enough organelles for two new cells. Copies of the DNA are also made so that each new cell will have its own set of chromosomes.

How many cell divisions occur in an adult human every 24 hours?

Mitosis begins after interphase. Mitosis is the stage in which the cell divides its nucleus in two. There are four basic phases in mitosis. During these phases the membrane around the nucleus disappears and the DNA condenses and coils into X-shaped chromosomes. The double set of chromosomes aligns in the middle of the cell. Then they pull apart, and one of each set moves to opposite sides of the cell. Two nuclear membranes form and then the chromosomes uncoil.

Once a cell has gone through mitosis, its nucleus and DNA have been doubled and then divided evenly. The rest of the cell must then split apart. Cytokinesis is the process in which the cell's cytoplasm is completely divided in half. In animal cells the cell membrane squeezes together, separating the two nuclei. The cell starts to look as if the cytoplasm is being pinched. In plant cells a cell plate first forms to separate the two nuclei, and then a cell wall forms.

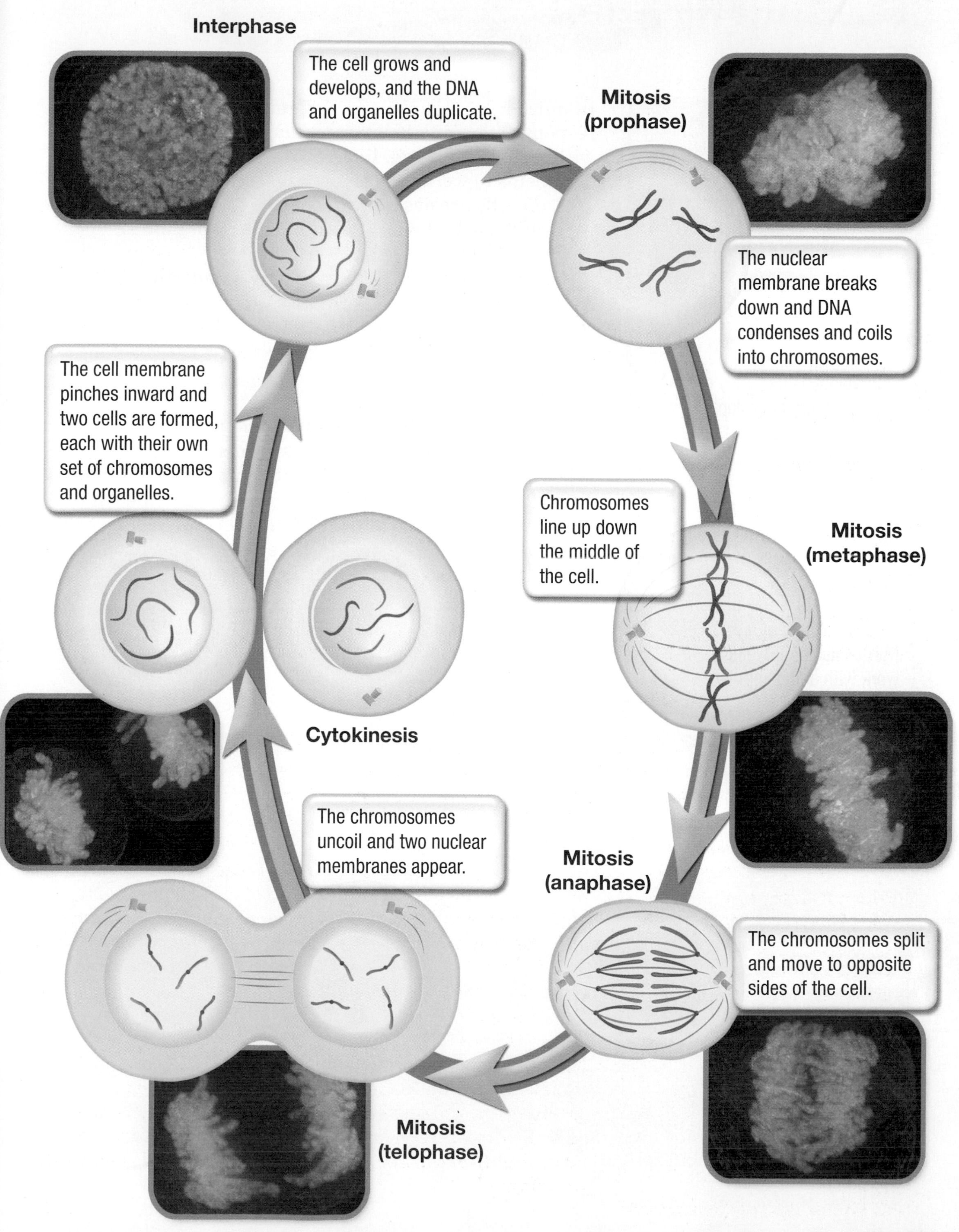
Interphase
The cell grows and develops, and the DNA and organelles duplicate.
Mitosis (prophase)
The nuclear membrane breaks down and DNA condenses and coils into chromosomes.
Chromosomes line up down the middle of the cell.
Mitosis (metaphase)
Mitosis (anaphase)
The chromosomes split and move to opposite sides of the cell.
Mitosis (telophase)
The chromosomes uncoil and two nuclear membranes appear.
Cytokinesis
The cell membrane pinches inward and two cells are formed, each with their own set of chromosomes and organelles.

CELL ORGANIZATION

Embryology is the study of how an embryo develops. Embryologists study the formation, growth, and development of living organisms. They also study abnormalities in embryos to better understand how and why they happen. Clinical embryologists study human embryo development. Non-clinical embryologists work with animal or plant embryos.

In unicellular organisms every function—growth, response to environment, reproduction—must occur within that single cell. In order for multicellular organisms to work properly, cells are often specialized. Certain cells are designed in such a way that they perform a specific function within a multicellular organism.

When a new organism begins to grow, the fertilized egg has all of the genetic material, or DNA, necessary for that organism to develop and survive. The new organism starts out as a single cell and begins to divide, forming an embryo.

Embryonic cells continue to divide and the cells form into layers. Animals with radial symmetry (a circular body type around a center) organize into an outer layer and an inner layer. Animals with bilateral symmetry (the left and right sides of the body mirror each other) produce a third, or middle layer. These layers will eventually develop into all of the animal's tissues and organs.

What happens when an egg is not fertilized?

When a new organism begins to develop, the single fertilized egg divides into two cells. These two cells continue to divide, forming an embryo.

Biologists have classified levels of organization in order to identify groups of cells that work together. In multicellular organisms the most basic level of function is an individual cell. Similar cells combine to form a tissue. Types of tissue include muscle, nerve, connective, and epithelial.

Several types of tissue that work together for a common purpose form an organ. For example, the human heart is made of large amounts of muscle tissue. However, there is also nerve tissue that stimulates the heart to move, as well as connective tissue that holds it in place. An organ system is formed from several organs that work together to perform specific tasks. Organ systems work together to form the organism. The organization of multicellular organisms is essential for their survival.

In the case of animals with bilateral symmetry, the inner layer will develop into organs like the lungs, liver, and pancreas. The middle layer will develop into various muscle and blood tissues. The outer layer produces the outer covering of skin as well as neurons in the brain.

OUTLOOK

Pain can be a good thing. While most people try to avoid pain, some people wish they actually could feel physical pain. Some people are born with a rare birth defect of the nervous system. It is referred to as *CIPA* (Congenital Insensitivity to Pain by Anhidrosis). People with CIPA cannot feel pain or distinguish between hot and cold. They often end up hurting themselves accidentally.

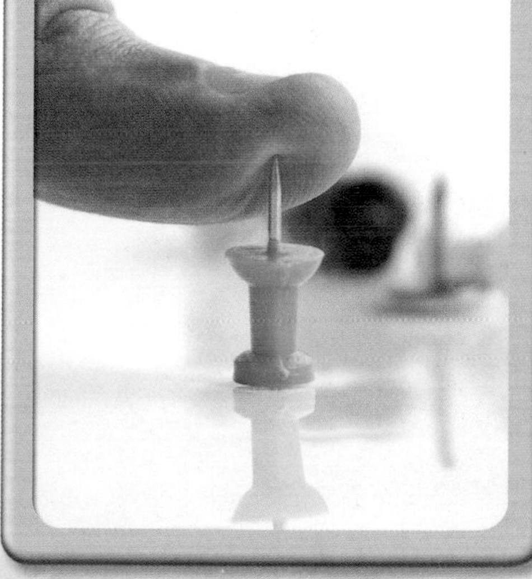

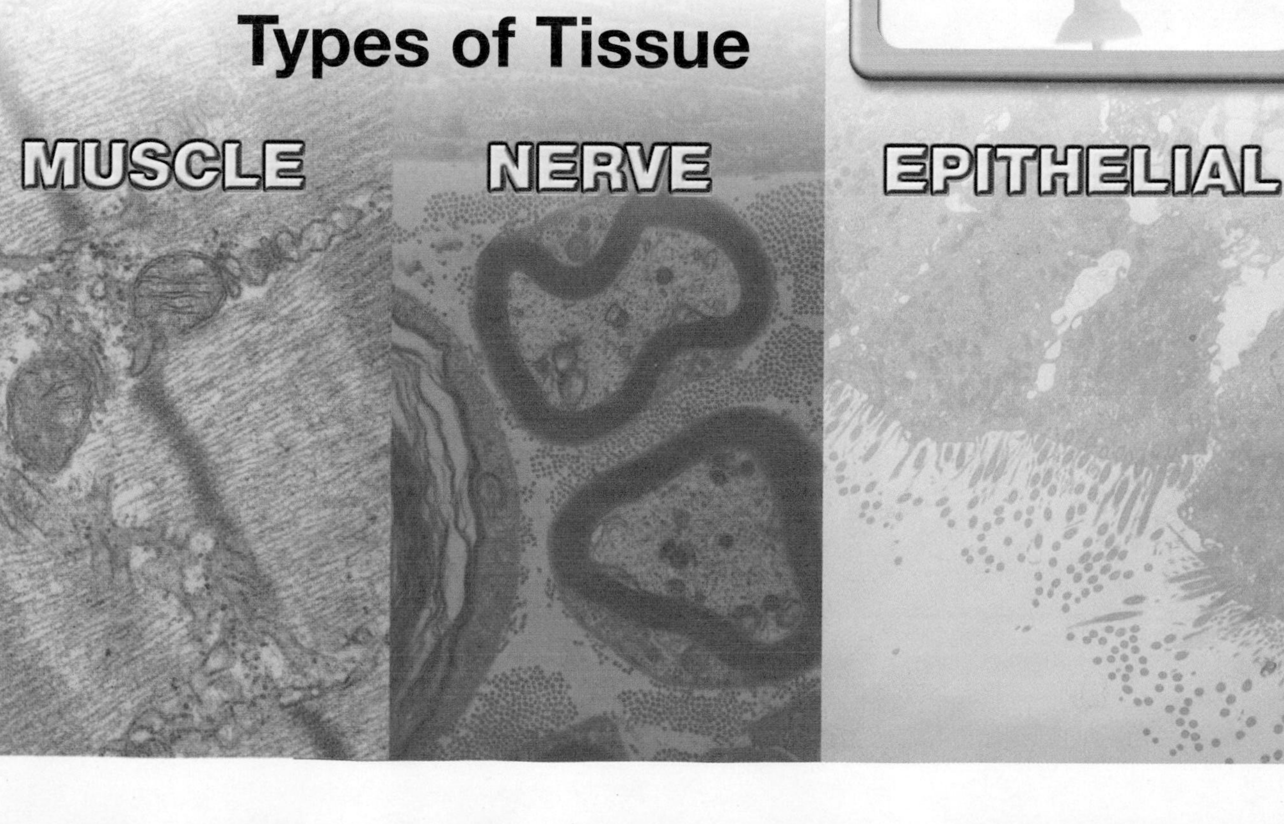

CANCER

VOCABULARY

mutation (myo͞o·ˈtā·shən) a change in the DNA sequence of a gene or chromosome

cancer (ˈkan·sər) a disease caused by cells that go through uncontrolled cell division

IN THE FIELD

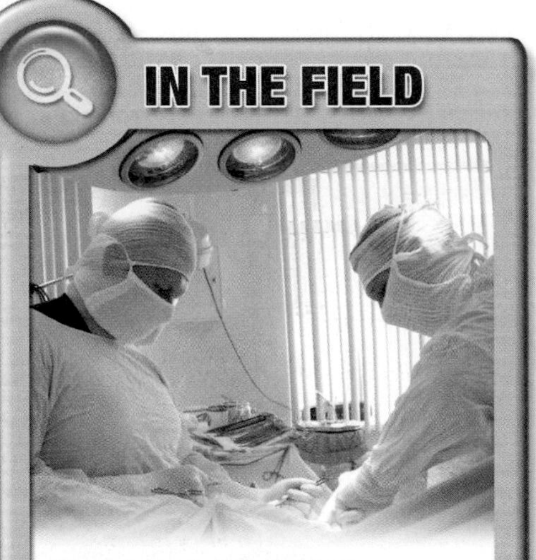

Oncology is the field of medicine that deals with the study, diagnosis, treatment, and cure of all different types of cancers. An oncologist is a medical doctor who has specialized training to help patients with cancer. From children to adults, this disease affects many people. While some forms of cancer can be treated and cured, scientists have not been able to find a cure for all of them.

Cell division is a normal process that occurs during the cell cycle that enables an organism to grow. Within a normal cell the process occurs a set number of times in a set amount of time. When a normal cell stops functioning properly, it dies.

Cells that go through the cell cycle divide in a controlled manner. Sometimes a portion of the DNA in a cell gets damaged. The damaged DNA causes a change, or mutation, in the cell. Since DNA carries all the instructions for cell functions, a mutation can cause the cells to function abnormally. Some mutated cells often grow and divide too fast. When this happens, it is called cancer. Cancer is a disease in animals and humans that begins when mutations cause cells to divide uncontrollably. It starts with one cell that develops in an abnormal way. The cell then divides again and again, producing more abnormal cells. This often results in a tumor, an illness, and sometimes, the death of the organism.

Do all mutations cause cancer?

Cancer cells develop abnormally. The yellow circular cells seen here are normal human cells. The rest are cancer cells.

Tumors are a mass of abnormal cells. They form when cancer cells divide and grow uncontrollably.

Cancer is a leading cause of death in many countries. There are over 100 different types of cancer, including leukemia, lung, liver, and skin cancer. Currently, only some forms of cancer can be cured. However, there are three common ways to treat it. One way is to remove the tumors through surgery. Chemicals may be used to try to destroy the cancer cells through a method called *chemotherapy*. Radiation therapy uses beams of high energy waves to damage large numbers of cells in order to keep them from reproducing. Sometimes these methods are used together to treat the cancer.

Eating a healthy diet with whole grains, fruits, and vegetables can lower the risk of some types of cancer developing. The overuse of tobacco products, alcohol, and other drugs can lead to lung, throat, and liver cancer. Sunscreen lotions with SPF 15 or higher help to block the sun's harmful rays, which can cause skin cancer. While cancer is not contagious, it is still important to try to reduce the chances of developing it.

LINKS

History Link
Research and write a report on the discoveries and contributions of Rosalind Franklin, Francis Crick, and James Watson.

Art Link
Draw and color the three stages of the cell cycle on a poster. Include the four phases of mitosis. Label each stage and phase and write a caption that describes what occurs in each one.

Journalism Link
Interview someone who has experienced having a type of cancer. Ask that person to share his or her story with you. In the style of a magazine article, write about what he or she experienced.

Skin cancer: A flat, red spot that has become rough, dry, and scaly over time may indicate skin cancer.

X-ray of lung cancer: A growth in the lung could possibly be lung cancer.

CHAPTER

4

ECOLOGICAL SUCCESSION

All communities change over time—sometimes quickly and sometimes slowly. A tree falling is a small change, but a forest fire or a volcanic eruption is a major event. Humans also bring about big changes in communities when mining or developing the land.

Soon after these interruptions occur, the area begins to renew itself. Scientists have discovered this process is orderly and predictable. The series of changes is referred to as ecological succession. Primary succession and secondary succession are the two basic types of ecological succession that can occur in a community.

VOCABULARY

ecological succession (ˌē·kə·ˈlä·ji·kəl sək·ˈse·shən) the predictable and orderly changes that occur over time in a community of plants and animals

KEY IDEAS

- Predictable and orderly changes, known as *ecological succession*, occur in communities.
- Ecological succession begins after a significant interruption takes place in a habitat.
- Interruptions are caused by natural events or human activity.

Communities and ecosystems, such as this beautiful one in Grand Teton National Park, Wyoming, go through a series of changes before they reach a mature stage. This natural process is called *ecological succession*.

QUICK FACT

The Hawaiian Islands were formed by volcanic eruptions. Today the islands have lush green jungles and a large variety of wildlife. However, this was not always the case. After the lava flowed and the islands began to form, the land was barren. Eventually plants began to grow in the cracks of cooled lava. Over time soil formed and conditions became more favorable to many new organisms. This series of changes formed the beautiful forests of the Hawaiian Islands.

Strange but True

Is this natural or man-made?

PRIMARY SUCCESSION

VOCABULARY

primary succession ('prī·mâr·ē sək·'se·shən) the series of changes occurring in a newly formed, barren habitat

substrate ('sub·strāt) the base, usually rock or soil, in which an organism lives

pioneer species (pī·ə·'nîr 'spē·shēz) the first species to populate or repopulate a barren or disturbed area

colonize ('kä·lə·nīz) to establish growth in an ecological community

ecology (i·'kä·lə·jē) the scientific study of the relationships between organisms and their environments

Primary succession is one of two types of ecological succession. It is more complex because it occurs in an environment in which new substrate has been exposed or deposited. It begins in an area in which there is usually no soil base. This type of succession can occur when a glacier retreats or a volcanic eruption leaves a layer of lava, creating or exposing bare rock. It can also begin on a newly-developed sand dune.

Primary succession begins when seeds, spores, or other organic materials arrive from nearby ecosystems and gather in cracks or crevices. If enough moisture is present, seedlings begin to grow. These first organisms that populate an area are called pioneer species. They often include moss, lichen, and weeds, which are especially suited to tolerate the harsh environment.

Pioneer species also include animals. Tiger beetles, wolf spiders, grasshoppers, and crickets are commonly the first animals to inhabit an area.

Why must plants and not animals be the very first pioneer species?

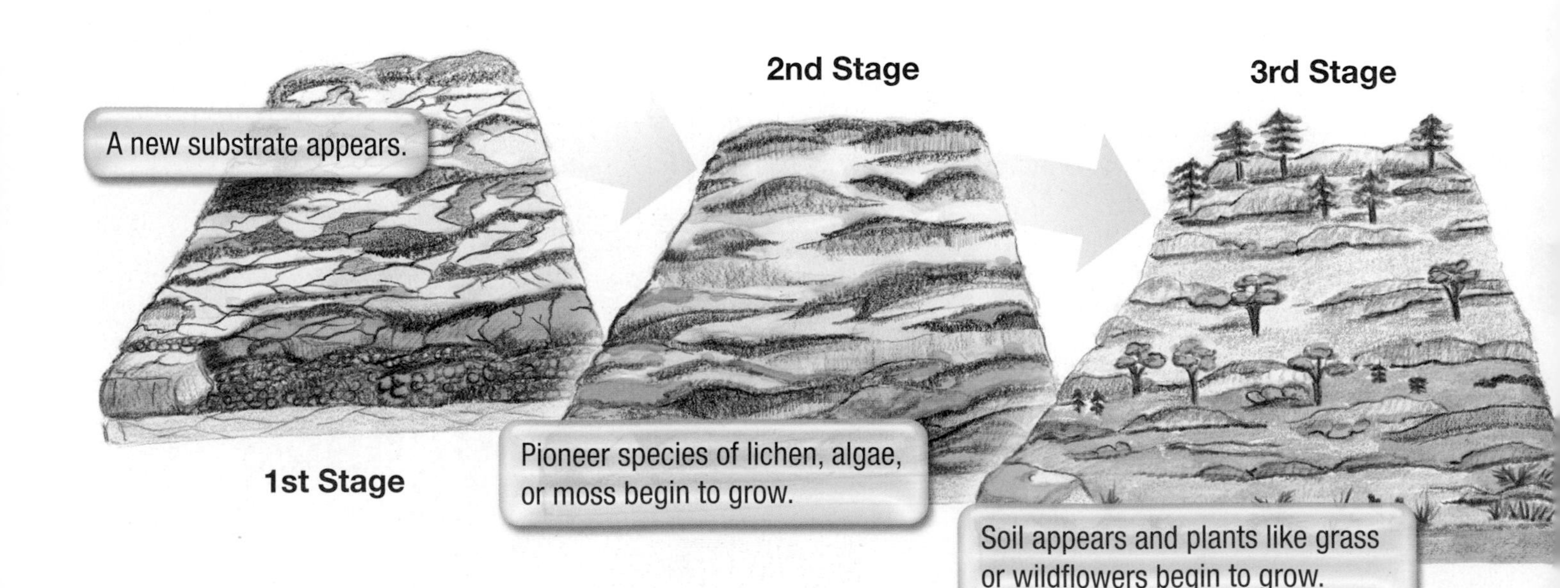

Lichen are complex organisms that produce acids to help break down rock. When these plants die, they decompose and provide organic material, which is needed for soil to form. When there is enough of a nutrient-rich soil base created, other species begin to appear. These are usually larger plants, such as shrubs, ferns, and grasses. They also add to the soil base, contributing the necessary nitrogen to the soil.

The types of plants in each stage of ecological succession depend upon the climate of the area and the plants in the surrounding ecosystems. As more varieties of plants colonize the area, a habitat that can support more animals begins to develop. As new species of animals populate the community, they bring in more plant seeds, adding to the variety of plant life.

The community continues to grow and flourish in this manner as it becomes mature and stabilizes. A disturbance would interrupt the balance of the ecosystem.

What do these burs and Velcro® brand hook and loop fasteners have in common?

4th Stage

A mature community grows in the fertile soil.

IN THE FIELD

Ecology is considered a branch of biology. It is the scientific study of the relationships between organisms and their environments. Ecologists examine these relationships by studying abiotic factors, such as rainfall, temperature, and sunlight. They also study the biotic factors that share the habitat. Because of the many different aspects included in the relationships between organisms and their environment, ecologists often draw on many other branches of science including geology, meteorology, genetics, chemistry, and physics.

Strange but True

This is pahoehoe lava from a volcano in Kilauea, Hawaii. The cracks and crevices create pockets ideal for lichen to begin growing.

GLACIER BAY

VOCABULARY

vegetation (ˌve·jə·ˈtā·shən) the plant life in an area

botanist (ˈbä·tə·nist) a scientist who studies plant life

sere (ˈsîr) a stage in the series of ecological succession

Glaciers scour the earth and leave nothing but bare rock as they retreat from an area. The absence of soil results in barren land lacking in vegetation and animals. This creates the condition for primary succession to occur.

One of the best examples of a glacial retreat followed by primary succession can be found in Glacier Bay, Alaska. John Muir was the first scientist to begin exploration of the area in 1879. At that time the glaciers had retreated approximately 48 km (30 mi) from where they were seen in 1794. William S. Cooper, a botanist, studied the plant succession when he arrived in 1916. Glacier Bay became a U.S. national park in 1980 and a biosphere reserve in 1986, protecting the area for future scientific study as well as public enjoyment. Today, glacial retreat continues, enabling scientists to further study the area.

QUICK FACT

Russia first offered to sell Alaska to the United States in 1859. The U.S. Civil War delayed the purchase. However, on March 30, 1867, Secretary of State William Seward offered Russia $7.2 million for Alaska. It officially became United States property on October 18, 1867. Many called Alaska *Seward's Folly* because they did not see any benefit to owning the land. Obviously that changed when gold was discovered!

What is the name of a boulder left behind by a retreating glacier?

When a glacier retreats, it leaves nothing but bare rock behind.

Black crust stabilizes silt and retains water. Its fibers make it feel and look somewhat like felt. Cyanobacteria, lichen, moss, green algae, microfungi and other bacteria form black crust. It is a pioneer species that helps create new soil.

Moss and horsetail begin to fill in the area once the soil is established.

The first types of vegetation that grew and began rebuilding the soil base were lichen and algae. They helped to break down the rock into soil and kept the soil base from blowing away. When the soil base began to develop, other plants like moss, horsetail, and wildflowers emerged. Then dryas and alders started to sprout and thrive. These plant species are important because they fix nitrogen into the soil. Wind, water, and animals brought the seeds for these plants.

As the soil became more fertile, a greater variety of vegetation repopulated the area. Shrubs, cottonwood, willow, spruce, and hemlock began to grow. Depending on the biotic and abiotic factors of each smaller area within Glacier Bay, salt marshes, bogs, meadows, or forests developed. Each stage of development, or sere, provided nutrients for the next sere.

Unlike the plants, there were no real pioneer species of animals to re-inhabit the area. Land animals must either walk or swim instead of being blown or carried in like seeds and spores. As a result, the repopulation of animal species has been a slower process.

Cottonwood

Spruce

What ingredient commonly found in kitchens is similar to the gray material pictured here?

Moss

Alder

Dryas

Forest dominated by spruce, hemlock, and cedar

Bog

SECONDARY SUCCESSION

VOCABULARY

secondary succession ('se·kən·dâr·ē sək·'se·shən) the series of changes occurring in an area where the existing ecosystem has been disturbed

softwood ('sȯft·wood) the wood from conifers or evergreen trees

hardwood ('härd·wood) the wood from broad-leaved, mostly deciduous trees

Sometimes hurricanes, floods, fires, or tornadoes interrupt the balance of an ecosystem. Certain human activities, like mining or the clearing of land for development, can cause a disruption too. Such events may upset, but not completely destroy, an ecosystem.

These types of disturbances can be minor or major. An example of something minor is a small landslide or a camper's fire being burned in an undesignated area. A major interruption could be a large forest fire, logging, or farming.

Disturbing an existing ecosystem in this manner causes secondary succession to begin. Secondary succession is the series of changes occurring in an area where an interruption has not totally destroyed the community. The soil base and even some organisms still exist. Therefore, this type of succession proceeds more rapidly than primary succession. Secondary succession happens in ecosystems ranging from ponds to large forests.

What started the 1988 fire in Yellowstone National Park?

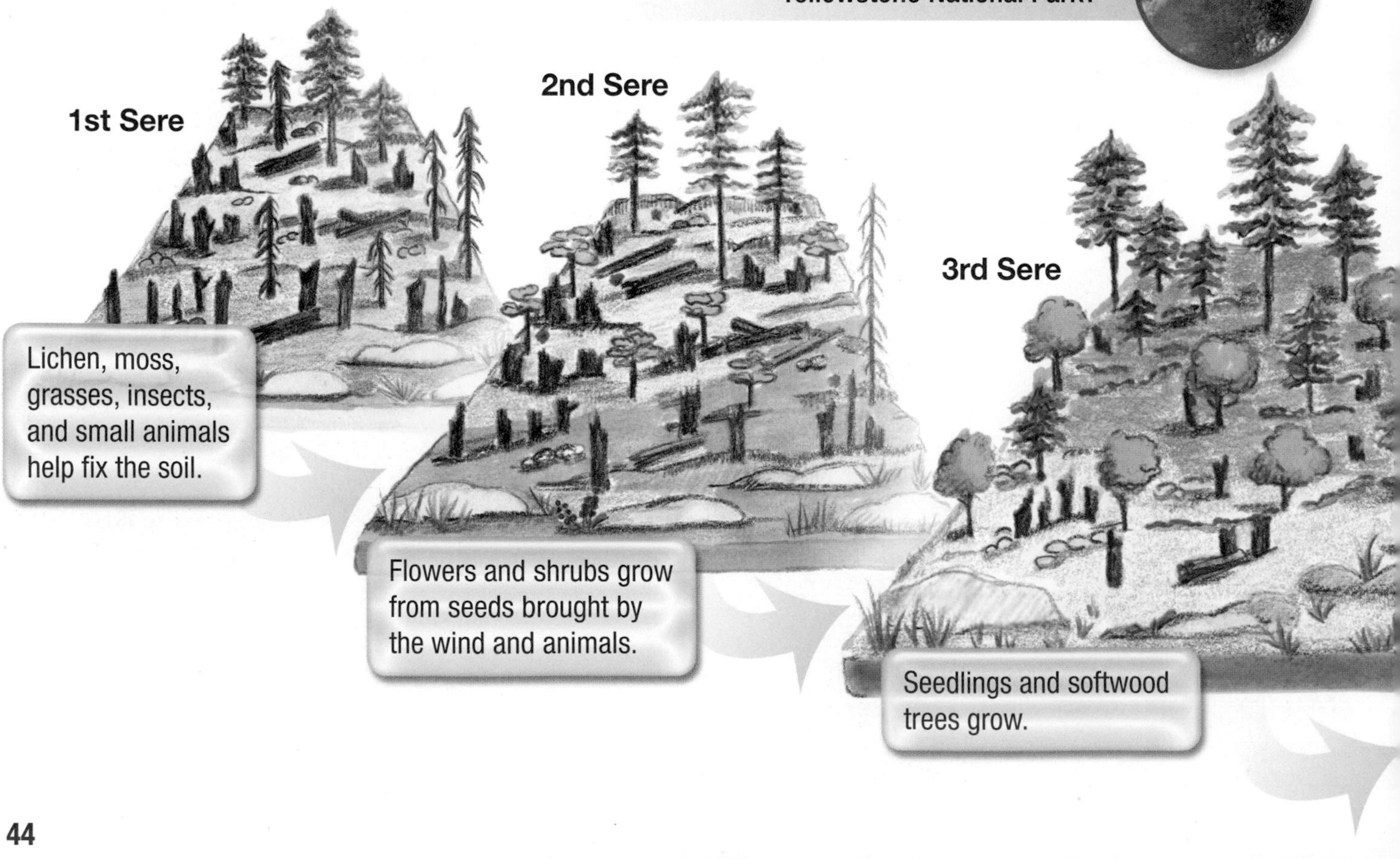

1st Sere

Lichen, moss, grasses, insects, and small animals help fix the soil.

2nd Sere

Flowers and shrubs grow from seeds brought by the wind and animals.

3rd Sere

Seedlings and softwood trees grow.

After a disturbance in a forest, the first sere yields various grasses, wildflowers, and mosses. These pioneer species, as well as insects and other small animals, help restore the soil.

Flowers and shrubs begin to flourish in the next sere. Various seeds are blown or brought in by migratory animals, and tree seedlings begin to grow. These plants begin to add nitrogen and other nutrients to the soil. Next, a forest may begin to develop. Pine tree seedlings continue to grow, eventually crowding out the grasses and flowers. This stage can take decades before the forest becomes mature.

Eventually, a mature, dense forest develops. Sometimes a forest consists of only one or two species of trees. At other times a forest may mature, having both softwood and hardwood trees. Softwood trees are conifers, such as pine and spruce. Hardwood trees, such as oak and elm, are deciduous trees that lose their leaves in the winter. A mature forest will dominate the landscape unless it is disturbed.

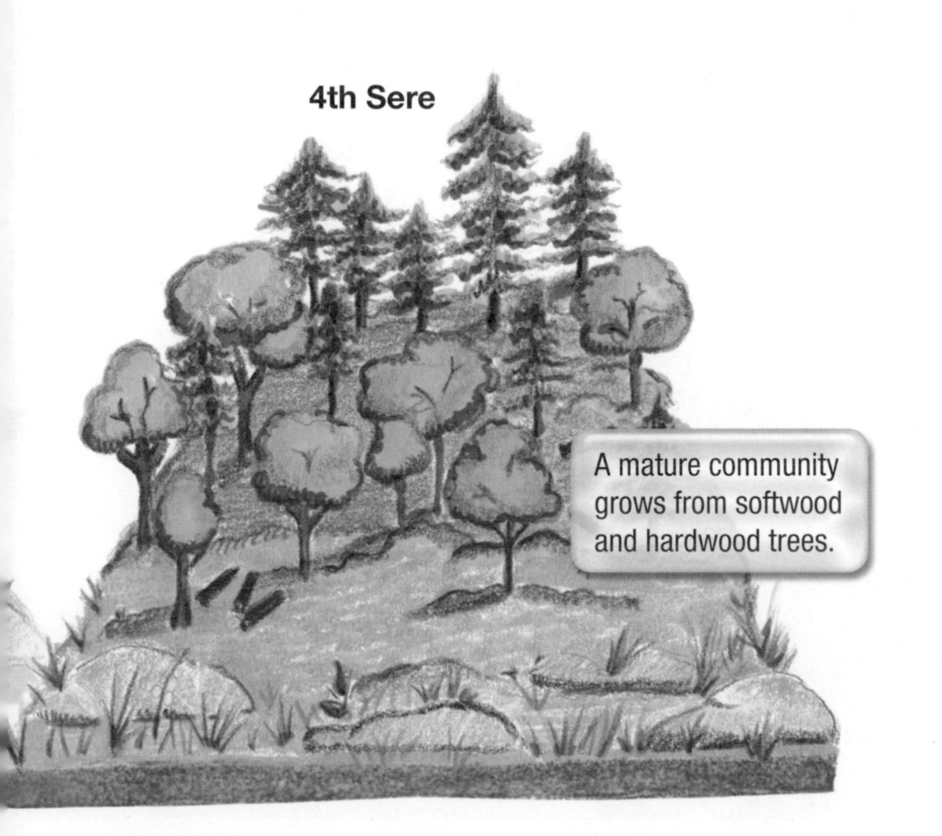

A mature community grows from softwood and hardwood trees.

QUICK FACT

Pines are softwood trees that have many uses. Did you know that turpentine, disinfectants, and artificial fragrances and flavorings are by-products of pine trees? Paints and adhesives contain rosin from the tree sap. It is what makes bandages sticky, gives baseball players better grips, and causes printing ink to bond to paper. Pitch is extracted from pine trees and then burned as fuel as well as added to medicines. Fatty acids that come from pines are used in fabric softeners, cleansers, and other materials. Pine trees benefit people greatly!

QUICK FACT

Even though a forest fire can destroy property and living things, it promotes ecosystem health as well. Fires are a natural way of clearing out thick underbrush. They also stimulate the regeneration of certain plants, such as sagebrush, aspen, and willow. The intense heat can crack open lodgepole pinecones, releasing the seeds held inside. Scientists have determined that the number and variety of plants and animal species increases if fires are allowed to burn out naturally.

The summer of 1988 was the driest that Yellowstone National Park had ever had up until that time. Its worst fire season followed. Heavy rains in April and May promoted large amounts of vegetation growth. June was extremely dry. By the end of July approximately 99,000 acres were ablaze. High winds and dry vegetation fueled the fires to make many of them uncontrollable. On August 20, 1988, there were 150,000 acres burning. September rain and snow helped extinguish the fires, but more than 793,000 acres, over one-third of the park, had been affected. More than half the acreage burned was from fires that had begun outside of the park.

There were 51 individual fires. Lightning caused 42 of them and humans started 9 others. The United States spent $120 million and some 25,000 people participated in the firefighting effort.

As many as 9,000 firefighters at a time worked diligently to help control or put out the fires.

Fires devastated the area, but did not destroy it completely.

Fires burned uncontrolled, scorching about 1.2 million acres in and around Yellowstone National Park.

While a number of plants and animals died, the fires did not severely damage the soil. In the years that followed, ample precipitation combined with ash and nutrients and made it possible for vegetation to recover quickly. Grasses and wildflowers replenished the area in just a few years. Aspen underground root systems were stimulated by the fire to begin new growth. Spruce, fir, and pine seedlings emerged. Many of the burned forests have been recolonized by lodgepole pines.

Overall, the animal populations were not greatly affected by the fires. Some of the grasses that elk eat are more nutritious now than before the fire. Scientists discovered that bears graze more frequently at burned sites than they do at unburned sites. Dead trees create more homes for cavity-nesting birds, such as bluebirds. No fire-related effects have been observed in the fish populations in the six rivers nearby. Life has regenerated the land in an amazingly short amount of time!

Grasses and wildflowers soon began to grow in the fire-blackened areas.

Pinecones released seeds because of the heat.

Various conifers began to recolonize burned-out forests.

New cycles of forest growth have caused the forest communities to recover quickly.

VOCABULARY

topography (tə·ˈpä·grə·fē) the physical surface features of a place or region

IN THE FIELD

Dr. Richard Oliver, also known as Ranger Rick, grew up in a godly home, but turned his back on Christianity as a teenager. He received his doctorate in evolutionary biology from University of California, Irvine. Dr. Oliver joined the first geological expedition after the Mount Saint Helens eruption. The changes he saw went totally against his beliefs. After seven years of research, he finally had to acknowledge there is a Creator. Ranger Rick now teaches around the world through his ministry, Confound the Wise, dedicated to spreading the unerring Word of God through science education.

Mount Saint Helens is a volcano located in southwestern Washington in the United States. On May 18, 1980, an earthquake with the magnitude of 5.1 occurred, triggering an eruption. Debris from the blast traveled up to 1,078 kph (670 mph) and devastated forestland. An ash cloud 19 km (12 mi) high deposited ash in states nearby. The area was devastated, but some spots were partly protected by the land's topography. Roots that survived and windblown seeds allowed plants to return quickly in these locations. Some regions were still covered with snow, which protected small trees and hibernating animals.

Areas that did not have deep ashfall were easily eroded by rain. Other places had such extensive damage that the succession was delayed because the soil was covered up to 183 m (200 yd) deep with ash and debris. Still in other sites, hot gases from the blast left the soil barren. In areas like these the soil needed to be exposed to air or enriched with nutrients before plants could grow successfully. These places took longer to begin recovery.

How much of the mountain's height was lost when Mount Saint Helens exploded?

Before the eruption Mount Saint Helens was considered one of the most beautiful and most frequently climbed mountains in the Cascade Mountain Range.

Pioneer plants, such as this fireweed, were the first to grow back.

Thousands of trees were blown down by the blast.

In March of 1989, the oil tanker *Exxon Valdez* ran aground, spilling 40 million L (11 million gal) of crude oil. Thick oil spread over the water and on the beaches, polluting 2,100 km (1,300 mi) of shoreline. Much of the oil spill affected Alaska's Prince William Sound, which is home to an abundance of wildlife. Because of the topography of the area, the oil remained contained in the sound, contaminating everything it touched. The oil also penetrated deep into the boulder beaches, prompting a massive cleanup effort over four summers. High-pressure water hoses were used to help remove the oil. Special bacteria were also introduced to help break up the oil.

In 2010 another massive United States oil spill occurred in the Gulf of Mexico off the coast of Lousiana. The Deepwater Horizon oil rig exploded and poured 750 million L (200 million gal) of oil into the water. To clean the oil spill, scientists used new technology and the knowledge gained from the *Exxon Valdez* cleanup efforts.

LINKS

Language Arts Link
Mount Saint Helens is part of the Ring of Fire. Research the seismic activity in this region and write a report about it.

Math Link
Develop a chart that tracks the total acreage burned during each week of a fire season in the area of your choice.

Art Link
Create a poster about fire safety to display in your school.

How would a killer whale be affected by an oil spill?

Birds, otters, sea lions, and many other types of sea life were affected by the spill.

It took almost 5 months to plug the Deepwater Horizon leak in the Gulf of Mexico.

The *Exxon Valdez* was carrying 200 million L (53 million gal) of crude oil. The disastrous oil spill led to the 1990 passing of the federal Oil Pollution Act.

Crews used high-pressure water washings to help remove the oil on the rocky Alaska shoreline. However, scientists soon recognized that this cleanup treatment killed off the surviving marine life and hindered the start of succession.

UNIT 1

LIFE SCIENCE: CYCLES

UNIT 2

PHYSICAL SCIENCE: TRANSFORMATIONS

– the study of the nature and properties of nonliving systems

UNIT 3

EARTH AND SPACE SCIENCE: PREDICTABILITY

UNIT 4

HUMAN BODY: BALANCE

CHAPTER 5 MEASURING MATTER

VOCABULARY

standard unit (ˈstan·dərd ˈyo͞o·nət) an established quantity used for measurement and comparison

KEY IDEAS

- Measurement compares an object to a standard.
- Many properties of matter can be measured.
- Measuring matter allows it to be precisely described.
- The advantages of the metric system make it useful to scientists.

One of the largest snakes in the world, the anaconda can attain a mass of 250 kg (551 lb) and grow to a length of 9 m (10 yd). How did scientists discover this information about anacondas? What makes an anaconda unique? The answers to both of these questions involve measurement. Measuring matter allows people to precisely describe and compare it. This is important in science. Without measuring an anaconda and comparing it to other snake species, its uniqueness might not be recognized.

Matter has many characteristics, such as length, area, volume, mass, temperature, and density. Many of these properties can change. In order to understand and clearly communicate about matter, it is measured before and after it is transformed. To help, scientists have developed standard units of measurement. A standard unit allows people to repeatedly determine the length of a road or the mass of a watermelon in the same way. Standard units enable people from different cultures who speak various languages to describe something in a way that is understandable to all.

OUTLOOK

In several passages of Scripture, God commands His people to use fair weights and measurements when buying and selling. (Leviticus 19:36, Proverbs 11:1, Micah 6:11) Israel's standard units and business practices were supposed to reflect the justice of God. What kinds of standards has God set for His people?

Strange but True

Would a trip to the moon make you lose weight?

VOCABULARY

milli- ('mi·li) a prefix meaning one-thousandth of a unit

centi- ('sen·ti) a prefix meaning one-hundredth of a unit

deci- ('de·si) a prefix meaning one-tenth of a unit

kilo- ('kē·lə) a prefix meaning one thousand units

Two major systems of measurement are currently in use around the world—the metric and the customary. Most nations rely on the International System of Units (SI), which the French developed. This system is also known as the *metric system*. Scientists worldwide use this system. England developed the customary system and introduced it to many of its colonies, including the United States. The United States and a few other countries use the customary system.

The metric system has some advantages over the customary system that make it practical for scientific measurements. The units of the metric system are based on the number 10. Multiples and fractions of 10 are relatively easy to work with, while many customary units have to be memorized. Calculations in the customary system are more difficult to do. The metric system uses a specific set of prefixes. The customary system does not use prefixes.

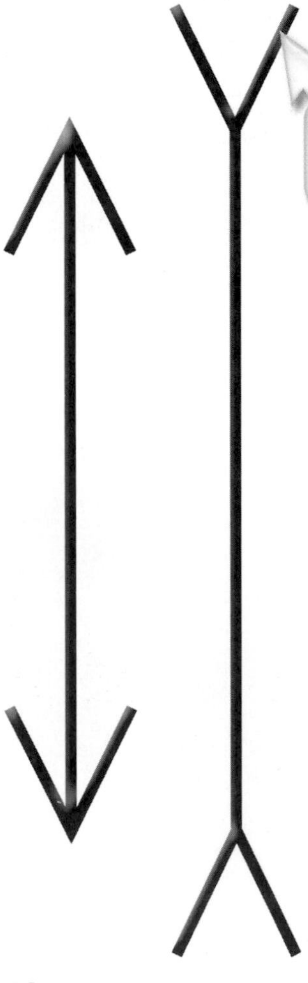

What measurements help to standardize this game?

Which line is longer? How can you tell?

In the game of cricket, the batsman uses a bat to hit a cricket ball. The length and mass of the bat, the volume and mass of the ball, and the distance between wickets (the three-legged stands behind the batsman) are all made according to standard metric measurements.

One characteristic or property of matter is length. The basic metric unit of length is the meter (m). A customary unit of length is the yard (yd). The chart below shows several units of length for both systems. In order to compare a metric length to a customary length, one must first be converted into the other.

The metric system uses several prefixes added to the basic unit, the meter, to indicate fractions or multiples of it. A millimeter (mm) is equal to $\frac{1}{1000}$ of a meter because milli- means $\frac{1}{1000}$. The prefix centi- means $\frac{1}{100}$ of a unit. When a meter is divided into 100 equal parts, each part is called a centimeter (cm). The prefix deci- means $\frac{1}{10}$ of a unit. Since a decimeter (dm) is $\frac{1}{10}$ of a meter, it takes 10 decimeters to equal the length of one meter. The prefix kilo- indicates that there are one thousand of a particular unit. A kilometer (km) is therefore equal to 1,000 meters (m).

QUICK FACT

In December of 1998 the *Mars Climate Orbiter* was launched and developed problems in 1999. Scientists hoped that this satellite would send back information about the climate on Mars. Part of the team involved in the project used customary units while another part of the team used metric units. No one realized the difference until it was too late. Instead of entering into an orbit around Mars as was intended, the *Mars Climate Orbiter* became lost in space.

How many centimeters are in one inch?

This car speedometer is labeled with both metric and customary units. The numbers closer to the middle of the speedometer are in kilometers per hour (kph). The numbers closer to the outside of the circle are in miles per hour (mph).

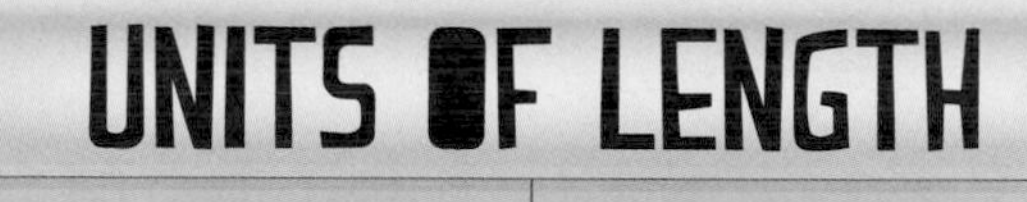

UNITS OF LENGTH

METRIC UNITS OF LENGTH	CUSTOMARY UNITS OF LENGTH
10 millimeters (mm) = 1 centimeter (cm)	12 inches (in.) = 1 foot (ft)
10 centimeters (cm) = 1 decimeter (dm)	3 feet (ft) = 1 yard (yd)
10 decimeters (dm) = 1 meter (m)	36 inches (in.) = 1 yard (yd)
100 centimeters (cm) = 1 meter (m)	5,280 feet (ft) = 1 mile (mi)
1,000 meters (m) = 1 kilometer (km)	1,760 yards (yd) = 1 mile (mi)

AREA AND VOLUME

VOCABULARY

volume (ˈväl·yəm) the amount of space matter occupies

Area and volume are measurable characteristics. Area is the number of square units required to cover a surface or region. To find the area of a rectangle, multiply its length by its width. For example, a 10 cm (4 in.) long and 5 cm (2 in.) wide rectangle has an area of 50 square centimeters (50 cm^2) or 8 square inches (8 $in.^2$).

There are plenty of practical reasons for measuring area. For instance, before painting a tennis court, one must calculate its area in order to estimate how much paint to buy. A standard tennis court is approximately 11 m (36 ft) wide and 24 m (79 ft) long. Multiplying 11 by 24 (or 36 by 79) reveals the number of square meters or feet that must be painted. Another example would be laying turf for an association football (soccer) field. Consider a playing field 91 m (100 yd) by 64 m (70 yd). The amount of artificial grass needed to cover this surface would be 5,824 square meters (m^2) or 7,000 square yards (yd^2). That is quite a bit of green!

Knowing how to find surface area can help you cover your school books.

Why is calculating area an important skill for an architect or construction worker?

1 linear centimeter

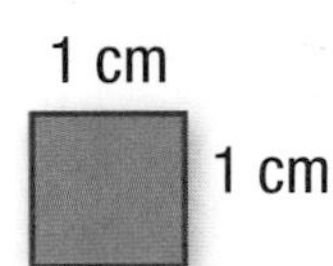

Area = 1 square centimeter = 1 cm^2

1 linear inch

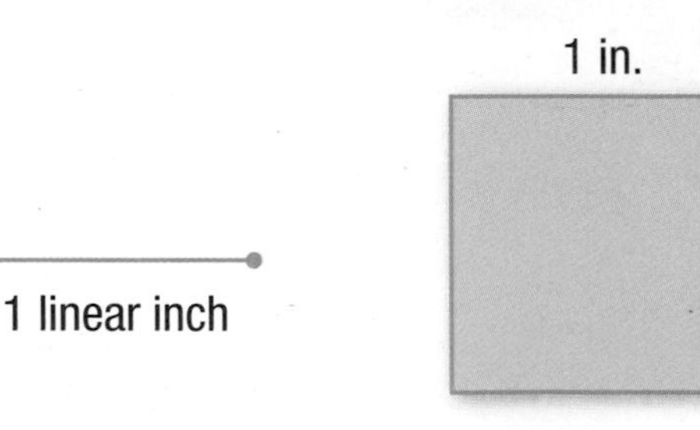

Area = 1 square inch = 1 $in.^2$

IN THE FIELD

Farmers use measurement and mathematics in their jobs. Farmers need to know the area of their fields to determine how much seed, fertilizer, and water to use. Many fields have an irregular shape, which can make calculating their area challenging.

Another way to determine the size of an object is to measure its volume. Volume is the number of cubic units within a three-dimensional object. Multiplying the object's length, width, and height provides the number of cubic units in it. Imagine a cube or rectangular prism with a length, width, and height that each measure one decimeter ($\frac{1}{10}$ of a meter). The volume would be one cubic decimeter (1 dm^3). Since there are 10 centimeters in each decimeter, the rectangular prism's volume is also 1,000 cubic centimeters (cm^3). The 1,000 cubic centimeters result from multiplying 10 cm by 10 cm by 10 cm. Interestingly, in metric measurement one cubic decimeter is also equal to one liter of volume. Therefore, volume can measure the amount of fluid in a container.

It is important to remember that all surfaces or regions have area and all three-dimensional objects have volume. Finding the area and volume of irregular shapes is not as easy as finding the area and volume of rectangles and cubes, but it is possible.

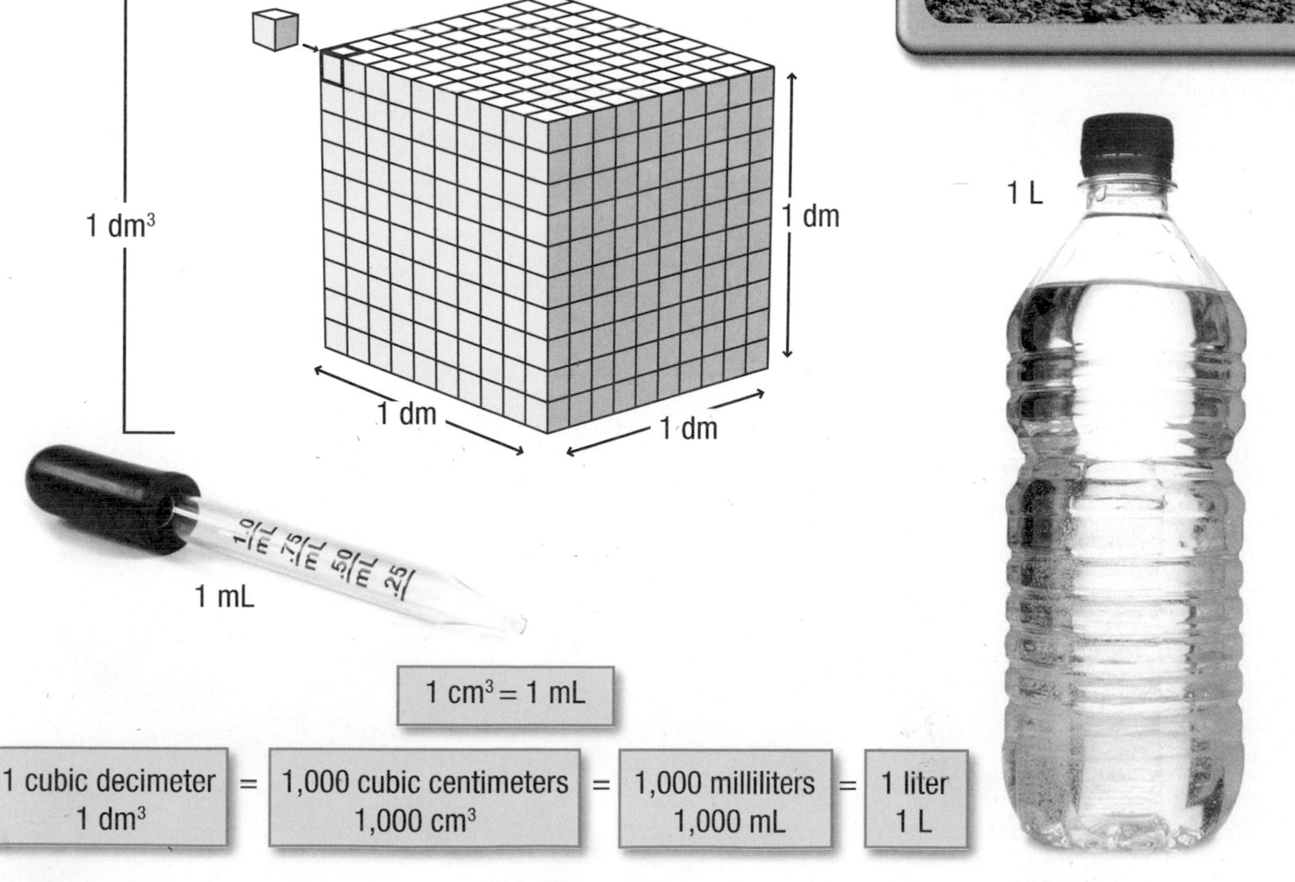

MASS AND TEMPERATURE

VOCABULARY

mass ('mas) a measure of the amount of matter of an object

weight ('wāt) a measure of the pull of gravity on an object

If you traveled to the moon, you would weigh one-sixth the amount of what you weigh on the surface of the earth. This is because the moon's gravity is six times less than the earth's gravity. However, your mass would remain exactly the same. Mass is not a measurement of the pull of gravity on an object. It is a measurement of the amount of matter of an object. This is one reason why scientists prefer to describe objects by using mass (kilograms) instead of weight (pounds).

A common way to describe an object is to measure its mass, or the amount of matter it has. The basic metric unit of mass is the kilogram (kg). One thousand grams equal one kilogram. Conveniently, in the metric system one milliliter (1 mL) of water has a mass of one gram (1 g). So, 1,000 milliliters (1,000 mL) of water, which is equal to one liter (1 L), has a mass of 1,000 grams (1,000 g) or one kilogram (1 kg).

The customary units of ounces and pounds are used to weigh an object. Weight is a measure of the pull of gravity on an object. An object's weight is not the same as its mass. Weight is dependent on the strength of gravity. Mass is not. Mass can be measured in kilograms and grams, but weight is measured in pounds and ounces.

Does the ant or the potato chip have a greater mass?

UNITS OF MASS/WEIGHT	
METRIC UNITS OF MASS	**CUSTOMARY UNITS OF WEIGHT**
1,000 milligrams (mg) = 1 gram (g) 1,000 grams (g) = 1 kilogram (kg)	16 ounces (oz) = 1 pound (lb) 2,000 pounds (lb) = 1 ton (T)
For water: 1 cm^3 = 1 gram = 1 milliliter	

Temperature is another attribute of matter. A person uses a thermometer to determine the temperature of something. Temperature is a measure of how much thermal energy an object has. The atoms and molecules of an object with a high temperature are moving faster than the atoms and molecules of an object with a low temperature. The object with the high temperature has more thermal energy than the one with the low temperature. If you touch a hot pan, thermal energy transfers from the hot pan to your cooler hand. This forces the particles in your hand to start moving faster. The faster these particles in your hand move, the hotter your hand feels.

The metric unit of temperature is the degree Celsius (°C). The customary unit for temperature is the degree Fahrenheit (°F). The freezing point of water is 0°C or 32°F. The boiling point of water is 100°C or 212°F.

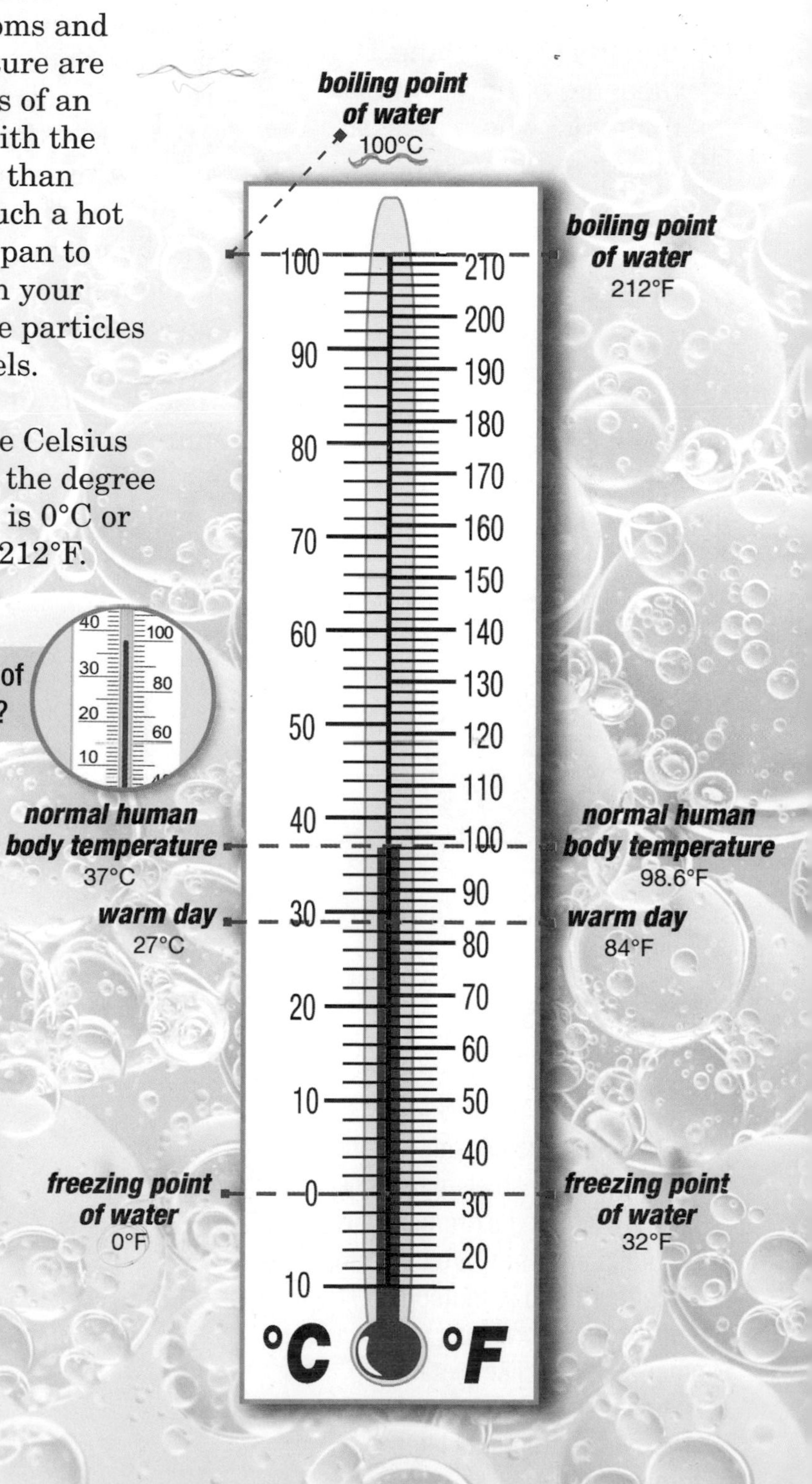

If this thermometer represents the temperature of water in a pan on the stove, is the water boiling?

When you are sick, your body temperature rises to help destroy harmful organisms in your body. This digital thermometer helps determine how high the child's fever is.

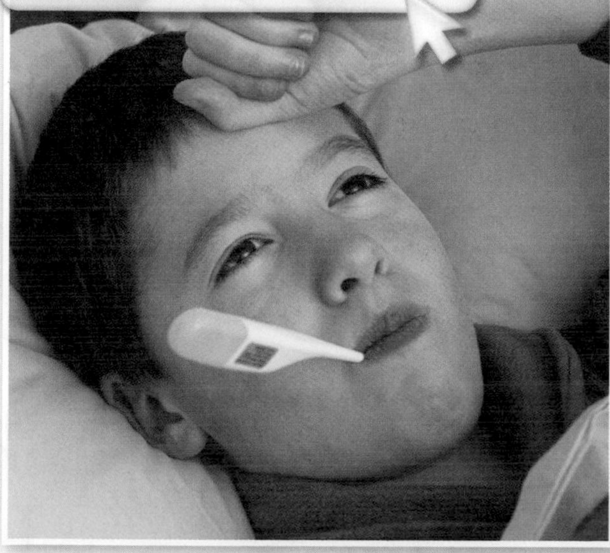

DENSITY

VOCABULARY

density ('den·sə·tē) the measure of how compact matter is, where mass is compared to volume

Certain properties of matter, such as an object's height or length, are easy to observe and measure. Other properties are not easy to see but can still be calculated. Density is the measure of how compact matter is or how much mass is contained in a certain volume of space. Imagine that you have two containers of liquid. The containers are exactly the same size and are filled to the top. One is filled with water and one with liquid cement. If you were to pick up both containers, the one with cement would feel heavier. Cement is more dense than water. Therefore, cement has more mass packed into the same amount of space when compared with water.

The density of matter is the same regardless of that matter's size. A piece of lead that you can hold in your hand has both a small mass and a small volume. The lead used in an X-ray apron has a larger mass and volume. However, because both objects are made of lead, they have the same density.

Why is a fishing line equipped with a plastic bobber?

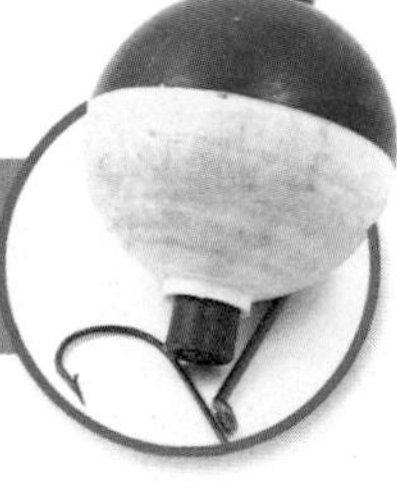

Before taking an X-ray of teeth, a dental technician puts an apron filled with lead over the patient's body. The density of lead is one property that allows it to shield the patient's vital organs from harmful X-rays.

Iridium, atomic number 77 on the periodic table, is considered one of the most dense elements. It is silvery white and extremely hard. Though somewhat rare on Earth, it is often found in meteorites. Iridium often combines with other elements like platinum and osmium. It is used to make precision instruments like surgical tools.

To calculate an object's density, divide its mass by its volume. $\frac{\text{mass}}{\text{volume}}$ = density.

For example, 1,000 mL of liquid water has a mass of 1,000 g. Its density is equal to 1,000 g (mass) divided by 1,000 mL (volume). $\frac{1{,}000 \text{ g}}{1{,}000 \text{ mL}} = 1$ g/mL

For a solid object, density is calculated in a similar way. A lead fishing sinker that weighs 330 g has a volume of about 30 cm^3. Notice that because the lead is a solid its volume is measured in cm^3 rather than in mL. Lead's density would be equal to 330 g (mass) divided by 30 cm^3 (volume). $\frac{330 \text{ g}}{30 \text{ cm}^3} = 11$ g/cm^3.

When you compare the two densities, lead is much denser, at 11 g/cm^3, than water, which is 1 g/mL. Lead sinks when dropped in water, not because it is heavier, but because it is denser.

QUICK FACT

Density affects both living and nonliving things. Ducks float in the water, but density affects how much they float or sink down in the water. Ducks have hollow bones that help the ducks float high above the water line. Loons and anhingas both have dense bones filled with marrow, so their bodies sit low or sink below the water line. The dense bones also help the birds dive deep into the water to catch fish.

Is air more dense at the top of a mountain or at sea level?

To keep a young child safe in water, an adult often places a flotation device around the arms. These plastic "wings" are filled with air. Since air is less dense than water, it makes the wings and the child float.

Helium is less dense than air, making it an ideal gas to fill up a blimp or zeppelin. Measuring the density of gases like helium requires special tools.

BUOYANT FORCE

VOCABULARY

fluid ('flo͞o·əd) a substance that can flow

buoyant force ('boi·ənt 'fôrs) an upward force in fluids that opposes gravity

displace (dis·'plās) to push aside

IN THE FIELD

In the United States, NASA constructed a pool of water called the *Neutral Buoyancy Laboratory*. It is used to train astronauts for the weightless condition of space. Scientists adjust the weight of the astronaut's equipment until the overall weight is equal to the buoyant force produced by the displaced water.

Gases are considered fluids because they can flow. Why do objects with less density float in more dense fluids like water and air? The earth's gravity pulls down on all objects. The only way to make an object float is to push up on it with a force equal to the object's weight. This upward force is called buoyant force or *buoyancy*. The buoyant force is equal to the weight of the fluid pushed aside. When a block of wood is placed in water, the wood pushes to the side, or displaces, some water. The weight of this displaced water is equal to the amount of buoyancy exerted on the wood. If the buoyancy of the wood equals the weight of the wood, it floats. If the wood is pushed under the water, the buoyant force becomes greater because more water is displaced, and the block floats back up.

What is sinking and what is rising in the water?

buoyancy

weight

This buoy floats because the upward buoyant force on it equals the downward force of gravity. Buoys are used for many purposes, including to help sailors navigate and to warn swimmers of danger.

When water seeps into wood and takes the place of air in the wood, the wood is called *waterlogged*. Waterlogged wood weighs more than the amount of water it displaces. Therefore, it sinks.

Buoyancy explains why huge steel ships float. Ships have a shape that displaces a large amount of water. When the weight of this displaced water equals the weight of the entire ship, the ship floats. Buoyancy pushes up on the ship and balances out the force of gravity, which pulls down on the ship. An important part of the construction of a ship is to make it contain a large amount of air. Air adds volume to a ship without adding much weight.

Submarines use buoyancy to control submerging and floating. Submarines contain large tanks called *ballast tanks*. Ballast tanks are filled with water to make the submarine submerge. When it is time for the submarine to rise, air is forced into the ballast tanks. The air pushes the water out. This makes the submarine lighter so that buoyancy can push the submarine up.

Would a hot air balloon float in Mars' atmosphere?

LINKS

Bible Link
Research the standard units of measurement that are mentioned in the Bible. See if you can discover the modern equivalent of a cubit. Use this information to describe the size of the ark mentioned in Genesis 6.

Art Link
Sketch a portrait of someone you know. To keep a figure proportional, artists use the length from the top of the head to the chin as a standard unit. The average human adult is about $7\frac{1}{2}$ heads tall. The armpits are about two heads from the top of the head.

History Link
Research and create a timeline for the metric system. Include key people and events that shaped its use around the world. You may choose to use drawings or a computer presentation of your findings.

Cranes are used to load and unload this cargo ship. Cargo is carefully weighed so that the weight of the ship and its cargo equals the weight of the water the ship displaces.

Submarines are used for scientific research, military purposes, and as tourist attractions.

CHAPTER 6

CHANGING MATTER

The physical part of the universe that has mass is called *matter*. The building blocks of all matter are elements, such as oxygen and carbon. An atom is the smallest unit of an element. All atoms of an element are the same type. When atoms join together, they become molecules. For example, when one atom of sodium combines with one atom of chlorine, a molecule of sodium chloride, or table salt (NaCl), is formed. Molecules that contain more than one type of element are known as *compounds*.

Atoms, elements, molecules, and compounds are types of substances. A substance is a single kind of matter that is pure. Each substance has its own unique set of properties which can be used to identify the substance. Physical properties, such as color and odor, are the characteristics of matter that can be observed or measured without changing the type of matter. A physical change involves altering some of these physical properties. Examples of physical change include molding clay, chopping wood, crushing aluminum cans, braiding hair, and making snow. In each example the type of substance remains the same. Only its form or size is altered.

VOCABULARY

substance ('sub·stəns) a single kind of matter that is pure and possesses a specific set of properties

physical property ('fi·zi·kəl 'prä·pər·tē) a characteristic of matter that can be observed or measured without changing the type of matter

physical change ('fi·zi·kəl 'chānj) an event that alters the form or size of matter but does not change the type of matter

Cl
Na
Na
Cl
Cl
Cl
Na
Na
Cl

This model of the compound sodium chloride represents a salt crystal magnified to the atomic level.

Strange but True

What kind of change is happening here?

KEY IDEAS

- Physical changes only alter the form or size of matter while the substances keep the same properties.
- Changing matter often involves breaking or forming various types of bonds.
- Chemical changes form new substances with different properties by rearranging the atoms within molecules.
- Nuclear changes form different elements by changing the nucleus of an atom.

IN THE FIELD

Employees at ski resorts observe and measure weather conditions. They wait for the right temperature and humidity to turn on the snowmaking machines. Snowmakers use compressed air and water pumped from nearby rivers. Inside the snowmaker the water turns into vapor before it condenses into snowflakes. In one hour snowmakers can create enough snow to cover an entire football field 1.7 m (3.5 ft) deep.

VOCABULARY

chemical bond ('ke·mi·kəl 'bänd) the force of attraction that holds atoms together within a molecule

In each type of substance, atoms are held together by positive and negative forces. Chemical bonds are the forces of attraction that hold the atoms together to form a molecule. In a water molecule, hydrogen and oxygen are held together by a chemical bond. These atoms are slightly positive or negative, even though they are all within a molecule. The positive end of one molecule attracts the negative end of another, just as opposite ends of magnets attract each other. This weak bond that holds two or more molecules together is known as *the van der Waals force*.

Changing matter involves changing the bonds that hold it together. One way to do this is to break the bonds between molecules. Van der Waals forces are the easiest to break because they are the weakest. When this type of bond is broken, a physical change occurs. For example, liquid chocolate hardens as tighter bonds form between the molecules. Also, when thermal energy is transferred to a chocolate bar in your pocket, the chocolate melts. The chemical bonds are broken and the molecules are able to move around.

Can physical change break down a compound into a new substance?

Physical changes took place to form, mold, and carve this sand castle.

Dutch physicist Johannes Diderik van der Waals (1837–1923) discovered that the state of matter is due to the forces between molecules. After finishing primary school, he worked as a teacher and attended university. In 1910 he won the Nobel Prize in Physics.

During a physical change, the big change is the distance between the molecules. They either get closer together or become farther apart. Suppose you took a piece of paper and tore it into so many pieces that it was hard to see the paper any more. The molecules that make up the paper are separated, but the paper is still paper.

Another type of physical change occurs when a compound like sugar dissolves in water. The water breaks apart, or dissolves, the bonds that hold the sugar molecules together. As the sugar molecules are dissolved, they become so small that they are invisible. Would the sugar disappear? Tasting the water proves it is sweet. If that water evaporates, a white solid is left in the bottom of the container. That is the sugar! The dissolving process simply alters the appearance, not the type of matter.

These salt deposits are in Death Valley National Park, United States. Where there used to be saltwater, now mostly salt remains because the water evaporated.

Can a physical change transform clay into this beautiful pottery?

This is a model of a molecule of table sugar, or sucrose ($C_{12}H_{22}O_{11}$). The blue spheres are carbon atoms. The white spheres are hydrogen. The red spheres are oxygen. The chemical bonds between the atoms are shown by the white connectors.

MIXTURES AND SOLUTIONS

VOCABULARY

mixture ('miks·chər) a combination of two or more substances that can be physically separated

solution (sə·'lo͞o·shən) a type of mixture of two or more substances that are evenly distributed

soluble ('säl·yə·bəl) able to be dissolved in a given solvent

insoluble (in·'säl·yə·bəl) unable to be dissolved in a given solvent

Another type of physical change is the formation or the separation of a mixture. A mixture is a combination of two or more substances which can be physically separated. Some mixtures are easily detectable, because you can see different parts. For example, a salad is a mixture of vegetables like lettuce, celery, and carrots. A pile of pebbles is a mixture of different rocks of various sizes, shapes, and colors. Other mixtures cannot be identified easily, such as air, soil, fruit juice, and cereal.

Separating a mixture into its individual parts requires a physical change. For instance, trail mix can be divided into groups of its ingredients. This alters only the size and appearance of the mix. The physical changes do not change the identity of its matter. The trail mix ingredients still taste and feel the same after they are no longer part of the mix. The same is true when a mixture of iron filings and sand is separated using a magnet. Both the iron and the sand keep their own individual properties.

Is the concrete in these blocks a substance or a mixture?

Physical change is required to separate a mixture.

This cup of hot chocolate is a mixture because it contains several ingredients, including chocolate powder, milk, and marshmallows.

A mixture is called a solution when the substances are evenly distributed. Solutions appear to be a single substance, because the individual parts are too small to see. Apple juice is a solution of sugar and other substances that are dissolved in water. In this case the water is the solvent because it is dissolving other substances.

When a substance, such as sugar, dissolves in water, it is said to be soluble in water. If you mix apple seeds with water, they settle to the bottom. The apple seeds are insoluble in water and will not dissolve, which creates a mixture, not a solution. Some paint spills can be cleaned up using water because they are water-based paints and have the ability to dissolve in water. Other paints are oil-based and are insoluble in water. To clean up oil-based paints, you must use a solvent like turpentine. The oil-based paint will dissolve because oil is soluble in turpentine.

What happens when water is used to clean up oil-based paints?

Sugar molecules are separated when dissolved in water. As the particles dissolve, they become too small to be seen with human eyes.

OUTLOOK

Is tap water as pure as it looks? No, tap water contains varying levels of dissolved chemicals and minerals, such as fluoride, chlorine, iron, manganese, and sometimes even gases. Soda water is made by dissolving carbon dioxide gas into a solution. Purified bottled water is not even pure because it, too, has substances dissolved in it. Pure water as a substance can be prepared in a laboratory. Although a substance may look pure, it could actually be a mixture.

Fruits, such as this lemon slice, have both soluble and insoluble components. The juice from the lemon is soluble in water. The rind, seeds, and inner membranes are insoluble.

CHEMICAL CHANGE

VOCABULARY

chemical change
('ke·mi·kəl 'chānj) an event that rearranges chemical bonds and forms one or more new substances with different properties

chemical property
('ke·mi·kəl 'prä·pər·tē) a characteristic that describes when or how a substance will interact with other substances

combustibility
(kəm·ˌbus·tə·'bi·lə·tē) the ability to burn

Matter can also undergo a chemical change by way of forming or breaking chemical bonds. During chemical changes the atoms within a molecule are rearranged to form new molecules or substances. Chemical bonds between atoms are stronger than bonds between molecules, so it takes more energy to produce a chemical change than a physical change. A burning match exhibits a chemical change. Chemical changes can also be called *chemical reactions*.

Chemical properties describe how or when a substance will interact in the presence of other substances. The range of chemical changes that a substance can undergo depends on its chemical properties. A chemical property known as combustibility describes a substance's ability to burn. In the presence of oxygen, both wood and gasoline are combustible.

How can you tell that a chemical change is occurring here?

Epoxy is a type of tough, strong glue made from compounds that are in resin and a hardener. A chemical reaction occurs and gives off heat when the compounds combine. Epoxies are used to join ceramics, copper pipes, wood, fiberglass, glass, and cloth.

Baking soda and vinegar are commonly used to make volcano projects because of their ability to react. Upon mixing the two substances, a third substance, carbon dioxide gas, is produced.

Remember that during a physical change, only the form or size of a substance is altered. The identity of the substance stays the same. In chemical changes, however, substances react and become completely new substances with different properties. Table salt is formed from the elements sodium and chlorine. Sodium is a shiny, silvery metal that conducts electricity, and pure chlorine is a yellowish-green gas that is poisonous to breathe. Yet, these two elements react to form a tiny, white crystal that is essential for life.

Chemical changes often cannot be reversed. For example, toasted bread cannot be cooled and changed back into untoasted bread. This evidence proves that a chemical change has occurred. Contrast toasting to the physical changes of melting and freezing. Frozen and melted water can easily be reversed, and the water will still be water.

QUICK FACT

Have you ever used a glow stick? When a glow stick is bent the first time, a glass vial inside breaks, releasing a chemical that initiates a chemical change. The chemical that was in the glass vial mixes with the chemical in the plastic tube. As the substances interact, their atoms become rearranged. Light energy is produced during the reaction, causing the glow stick to glow. After both chemicals have reacted, the glow stick stops producing light.

Sodium and chlorine in element form have completely different physical properties than the compound sodium chloride. Sodium is combustible in water, producing bright flashes of light. Chlorine was used in World War I as a poisonous gas. Nevertheless, sodium and chlorine combine to form table salt.

NaCl

Na

Cl

An apple turns brown because it reacts with oxygen, or oxidizes. The chemical bonds inside of the apple's molecules change. Simply washing the apple does not return it back to its original state.

NUCLEAR CHANGE

VOCABULARY

nuclear change ('nōō·klē·ər 'chānj) an event that greatly alters the nucleus of an atom

Atoms are made of protons, neutrons, and electrons. Scientists call the center of the atom *the nucleus*. Protons and neutrons are found inside the nucleus, while electrons move rapidly around it. A chemical change involves only the electrons of atoms. The protons and neutrons are held together by a bond known as *the strong force*. A nuclear change will affect this force. During a nuclear change, an element can gain or lose protons, neutrons, and other particles within the nucleus. Nuclear changes convert elements into one or more new elements.

Nuclear changes often occur by fission or fusion. In the process called *fission*, the nucleus of a large atom splits into two smaller atoms. The radioactive element uranium is used to produce electricity during nuclear fission reactions. The second process, known as *fusion*, occurs when two or more smaller atoms combine, or fuse, into one larger atom. Fusion takes place within the sun and other stars when hydrogen atoms fuse together to form larger atoms, such as helium.

This nuclear power plant produces electricity through fission reactions. The energy released through this process is used to heat water into steam, which turns a turbine. The white cloud leaving the large cooling tower is mostly water.

What does this symbol mean?

This is an early model of the atom. The blue and green spheres in the middle represent protons and neutrons held together in the nucleus. The red spheres represent electrons. The nucleus of an atom transforms during nuclear reactions. Only the electrons are involved in chemical reactions.

God has designed several nuclear changes to occur naturally, such as the fusion that takes place within stars. In addition, the nucleus of certain elements spontaneously breaks apart and decays. Whenever any of these nuclear changes occur, a new element is formed. A typical nuclear reaction gives off nuclear radiation as a by-product. Sunlight is a by-product of the fusion reactions that take place deep within the sun.

Nuclear radiation consists of invisible rays similar to X-rays that can be both helpful and harmful. The radiation given off by the element americium has a beneficial role in smoke detectors. Many medical procedures use nuclear radiation to diagnose and treat diseases. Scientists work with radioactive elements to study plant processes. Engineers depend on radiation to detect flaws or leaks in metal. However, too much exposure to certain types of radiation can also cause cancer, kidney failure, emphysema, and other serious illnesses.

OUTLOOK

Deep in the earth's crust there is no light, little to no oxygen, high temperature, and very little water. For many years scientists thought that nothing could live under these extreme conditions. However, bacteria were found to exist in rocks as deep as 3.5 km (2 mi) below the earth's surface! Their energy source appears to be the radioactive decay of rocks like uranium. Amazing and all-powerful, God can even create organisms able to survive in rocks.

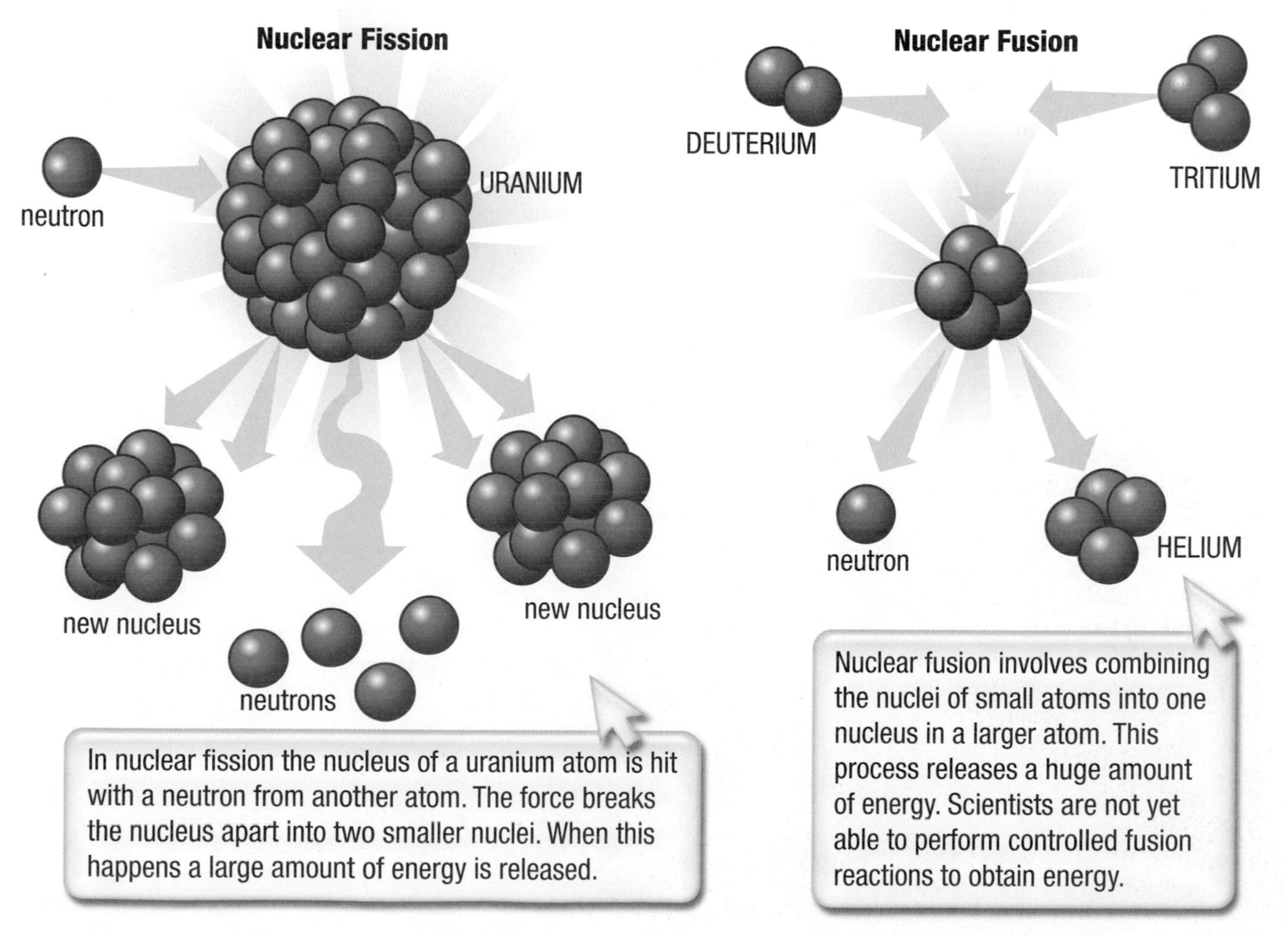

In nuclear fission the nucleus of a uranium atom is hit with a neutron from another atom. The force breaks the nucleus apart into two smaller nuclei. When this happens a large amount of energy is released.

Nuclear fusion involves combining the nuclei of small atoms into one nucleus in a larger atom. This process releases a huge amount of energy. Scientists are not yet able to perform controlled fusion reactions to obtain energy.

COMBUSTIBILITY

IN THE FIELD

Automotive designers and scientists work together with government agencies and auto makers to develop cars that produce less pollution. One alternative is an electric car that uses a hydrogen fuel cell, which frees the electrons to move around and create electricity to power the car. However, hydrogen is very combustible. For safety reasons, the fuel cell keeps the hydrogen and oxygen in separate cells. A regular car's fuel is gasoline that combusts and produces exhaust. A hydrogen car's fuel combusts and produces water (H_2O) in the form of steam.

Any substance that is combustible has the ability to burn. When a fuel burns, the chemical process of combustion takes place. During the process of combustion, the fuel reacts with oxygen, giving off heat energy. This heat energy can be used to accelerate objects. New substances are formed as a result of the chemical changes that occur during combustion. These new substances are called *exhaust*. Exhaust does not always harm the environment.

Hydrogen and oxygen are combustible. They are used as fuel to push rockets into outer space. The thrust comes from the fuel burning during combustion. Liquid hydrogen and oxygen are stored in a fuel tank attached to the space shuttle. The shuttle's three engines burn the hydrogen and oxygen, producing water vapor. The water vapor is released as exhaust, increasing the rocket's thrust.

fuel tank

rocket boosters

engines

Every 10 seconds during lift off, the amount of liquid hydrogen and oxygen burned could fill a family-sized swimming pool!

Every star, including the sun, produces an incredible amount of energy through nuclear change. Scientists estimate that hydrogen atoms must be heated to over 100,000,000°C (212,000,000°F) for fusion to occur!

Fossil fuels, such as coal or oil, also release heat energy during combustion. This heat energy can also make objects move. Many vehicles burn fossil fuels in a combustion engine. Combustion within the engine is what makes the engine parts move, which in turn moves a car. The exhaust leaves through the tailpipe of the car.

Nitrous oxide is a chemical used in combustion to increase the amount of available oxygen. This makes fuel burn faster. The more rapidly a fuel burns the more power a vehicle has. Rockets and race cars go faster and farther when they use nitrous oxide. Typically, air is only 21% oxygen. Nitrous oxide changes the air content to 33% oxygen. This makes combustion faster and more efficient.

A nitrous oxide kit installed in a race car increases the car's power and speed.

LINKS

History Link
Research the 1986 Chernobyl, Russia and the 2011 Fukushima, Japan nuclear disasters. Write a report about the causes of the disasters and how they affected the areas around them. Tell what improvements nuclear plants have made to prevent similar accidents in the future.

Math Link
If four hydrogen atoms are required to produce one helium atom through nuclear fusion, calculate how many helium atoms could be produced if there were 176,540 hydrogen atoms. Then calculate how many hydrogen atoms are required to produce 56,000 helium atoms. Show your work.

Technology Link
Research how industry uses nuclear radiation. Include smoke detectors, food irradiation, and any safety concerns regarding nuclear radiation. Write a report or make a computer presentation.

Bible Link
Read and spend time reflecting on Isaiah 60:19. Prepare a devotional about this verse and relate it to what you have learned about how God created the sun to produce energy.

CHAPTER 7

FORCE AND WORK

Forces push and pull objects. A balanced force occurs when the individual forces are equal in strength and opposite in direction. When these opposing forces act on an object, the object's motion or position remains unchanged. In other words, there is no change. However, an unbalanced force takes place when separate forces are not equal in strength or they are not opposite in direction. If the object is already moving, the unbalanced force causes a change in the object's direction. If the object is at rest, the unbalanced force will cause the object to move. A fixed reference point is used to determine motion. Without a reference point, it is difficult to see when an object moves.

KEY IDEAS

- Unbalanced forces affect motion.
- Motion can be described and measured by speed, velocity, and acceleration.
- Work occurs when an object moves in the same direction as the applied force.
- People use simple machines to make work easier.

Strange but True

The grasshopper's jumping action resembles what simple tool?

IN THE FIELD

In order for a rocket to lift off and to leave the earth's atmosphere, it must overcome the gravitational force of the earth. To do this, the rocket must exert a force greater than that of gravity. To generate this unbalanced force, the rocket must produce enough thrust to create a speed of about 40,000 kph (25,000 mph). Once this speed is reached, the rocket has enough power to leave the earth and its atmosphere.

This car was in motion until it became stuck in the sand. While the car is at rest, the forces acting on it are balanced. When the men push the car, they create a force greater than the balanced force, which holds the car in the sand. The unbalanced force of the men's push will move the car.

VOCABULARY

speed ('spēd) a measure of the distance an object moves in a given amount of time

velocity (və·'lä·sə·tē) a measure of an object's speed and direction

acceleration (ik·ˌse·lə·'rā·shən) the change in an object's velocity over a period of time

QUICK FACT

The cheetah is the fastest animal on land. It can reach speeds of up to 112 kph (70 mph). The fastest creature on Earth is actually the peregrine falcon. When diving, this magnificent bird may fly faster than 440 kph (273 mph)! Peregrine falcons are raptors, a kind of bird that hunts and kills prey for food. These falcons are easily distinguishable from other raptors because of the black feathers on their heads. Peregrines mate for life, are naturally calm, and easily tamed by humans.

When an object moves, the distance it travels can be calculated by measuring its position compared to a reference point. For example, the starting line for a race would be the reference point. The finish line would be the ending position for the racers. The length between the two lines is the distance the racer must travel.

One way to describe motion is by the distance an object travels in a given amount of time. Speed is a measure of how fast an object moves from one position to another. It is measured in length units per time units. The units are not always the same. Common distance units are meters (m), yards (yd), kilometers (km), and miles (mi). Units of time are typically seconds (sec), minutes (min), and hours (hr). To calculate speed, divide the distance an object moved by the time it took to reach that distance.

What is the flight speed of a hummingbird?

To measure this snowboarder's speed during the race, you would divide the distance she traveled by the amount of time it took for her to cross the finish line.

Calculations of Speed

Speed = distance ÷ time

$$\frac{2{,}000 \text{ m (meters)}}{40 \text{ sec (seconds)}} = 50 \text{ m/s (meters per second)}$$

$$\frac{2{,}182 \text{ yd (yards)}}{40 \text{ sec (seconds)}} = 55 \text{ yd/s (yards per second)}$$

Velocity is another way of describing motion. While speed is the rate of change in distance over time, velocity refers to both an object's speed and direction. Two objects may be moving at the same speed, but if they are moving in different directions, their velocities are also different. For example, a bus traveling east at 65 kph (40 mph) has a different velocity than a bus heading south at the same speed.

Acceleration is also used to describe motion. Any change in an object's velocity is considered acceleration. Usually people think of speeding up as acceleration and slowing down as deceleration. However, the scientific term for acceleration means any change in velocity—whether speeding up, slowing down, stopping, or changing direction. Therefore, acceleration describes the rate at which velocity changes.

IN THE FIELD

Speed skating, or racing on ice skates, has been taking place since the nineteenth century. Speed skating is a sport in which competitors are timed while traveling the length of a set track. It became an Olympic sport in 1924. A standard track is 400 m (437 yd) long, and skaters reach speeds as high as 60 kph (37 mph) during shorter distance races. Women were officially accepted into this Olympic competition in 1936.

A bicycle ride is a good example of acceleration. When the rider speeds up, slows down, or stops, acceleration occurs. When he turns in any direction, the acceleration changes also. The acceleration equals zero when the rider keeps the same speed and heads in the same direction.

These two boats have different velocities because they are traveling in opposite directions at different speeds.

WORK

VOCABULARY

newton ('nōō·tən) a metric unit used to measure force

work ('wûrk) the result of a force applied to an object that causes it to move

energy ('e·nər·jē) the ability to do work

joule ('jōōl) a metric unit used to measure work, equal to one newton meter (Nm)

Measurements provide data to analyze. Conclusions can then be made from the data. Sometimes it is necessary to know how much force is acting on an object. Newton's second law of motion states that force is equal to an object's mass multiplied by its rate of acceleration.

The greater the force applied to a mass, the greater the acceleration will be. For example, when an adult pushes on a box full of books, the acceleration is more than when a child pushes on the same box. Why? Because the adult is applying more force. If the same child were to push a heavier box of books, the acceleration would be even less because the mass is greater.

The relationship between force, mass, and acceleration is expressed in a mathematical formula. The unit used to measure force is a newton.

$$\text{force} = \text{mass} \times \text{acceleration}$$
$$f = m \times a$$

A fifth-grade student pushes an empty go-cart and causes it to accelerate.

If the fifth grader applies the same amount of force on the go-cart with someone sitting in it, the acceleration will be less than before. How can the student get the go-cart to accelerate more?

QUICK FACT

The newton is named after Sir Isaac Newton (1642–1727), one of the most influential Christians in the history of science. He was the first person to understand and explain the relationship between mass, acceleration, and force.

An unbalanced force can cause movement. Scientifically speaking, work occurs when an object moves in the same direction as the force being applied. If you lift a backpack off the floor, you are doing work because the force you are applying is upward and the backpack also moves upward. If you carry the backpack across the room, no work is done because the movement and the force are not in the same direction.

Work occurs when energy is used to apply force to an object and it moves in the direction of the force. When greater force is applied, more work takes place. In addition, more work is accomplished when an equal amount of force is applied but the object travels a longer distance. If force or distance increases, then work also increases. The amount of force times the distance that the object moves equals the amount of work. The unit used to measure work is joules.

$$\text{work} = \text{force} \times \text{distance}$$
$$w = f \times d$$

What takes more work, walking or running up a flight of stairs?

A hydraulic lift can generate enough force to lift a car. Work is done because the force applied is in the same direction that the car moved.

work	=	force	× distance
12,000 joules	=	6,000 newtons	× 2 meters

OUTLOOK

God created the muscles in the body to do work. You may think of work as something hard or unpleasant to do. But work can actually be fun and relaxing. In fact, many of the enjoyable things people do involve work. Hiking up a mountain and kicking a soccer ball both require applying a force. These actions are considered work because an object is moving in the same direction as the force.

When the force exerted by this man's arm brings the bowling ball backward and then forward, work is performed.

SIMPLE MACHINES

VOCABULARY

simple machine (ˈsim·pəl mə·ˈshēn) a device that makes work easier and which has few or no moving parts

lever (ˈle·vər) a beam or bar that pivots on a fixed point

inclined plane (in·ˈklīnd ˈplān) a straight, slanted surface used to raise an object

fulcrum (ˈful·krəm) the point where a lever rotates or pivots

Simple machines help make work easier. However, simple machines do not reduce the amount of work that is done. Work takes place when a force is applied to an object and that object moves. The simple machines either change the amount of force applied, alter the direction of the force, or both. Can you lift a couch onto a truck by yourself? Can you raise a car from the ground while someone changes a flat tire? You could do both of these tasks by using simple machines.

When enough force cannot be applied to move an object, certain simple machines help to do more work with less force. A lever is made of a beam or bar that rotates or pivots around a fixed point. An inclined plane is a slanted surface, such as a ramp, used to raise an object.

This water well uses a kind of simple machine called the *wheel and axle* to draw up a bucket of water. A wheel and axle is actually a type of lever.

Ramps are inclined planes that reduce the amount of force needed to move an object.

The handle of the hammer is a lever. It causes the head of the hammer to pivot and force the nail loose.

Car jacks, wheelbarrows, screwdrivers, wheels, and wrenches are levers. Even the human arm acts as a lever. A lever works when you apply a force to one part of the lever and it applies a force to an object at another part of the lever. Each lever pivots around a point called the fulcrum. In the case of a wheel, the fulcrum is also called *an axle*. You can turn the wheel to spin the axle or you can turn the axle to spin the wheel.

Ramps, wedges, and screws are inclined planes. Ramps decrease the amount of force required to move an object. A wedge is two inclined planes, which together push in opposite directions—like an ax splitting wood. A screw is a tiny inclined plane wrapped around a cylinder.

QUICK FACT

A pulley is another type of lever. It is a wheel with a rope around it, and people use it to change the direction of a force. Not as much force has to be applied when using a pulley, but the end of the rope must move farther than the load. Flagpoles, curtain rods, mini-blinds, and cranes are all examples of items with pulleys. They help to move things up, down, or across. Pulleys do not reduce the amount of work, but they do make work easier.

The edge of this axhead needs to be sharpened on both sides, because it has two inclined planes back to back.

This crane makes use of a lever. The point at which it is attached and pivots is the fulcrum.

A hand plane has an inclined plane with one horizontal side and one angled side and works like a blade. As it is pushed across the wood, it shaves off rough places and makes it smooth.

TYPES OF LEVERS

VOCABULARY

effort ('e·fərt) the force that is applied to a simple machine

load ('lōd) an object that resists the force applied by a simple machine

Levers are everywhere. Imagine you and a friend are on a seesaw. When you push down on one end, you are applying force that pushes the other end upward. This force that you apply to the lever is called the effort. The lever is doing work to lift your friend by changing the direction of the force. The weight of your friend represents the load.

There are three types, or classes, of levers. The first-class lever has the fulcrum between the effort and the load. Pushing down on the effort side lifts the side with the load. The second-class lever has the load between the effort and the fulcrum. Pulling upward on the effort side lifts the load while pivoting on the fulcrum. The third-class lever has the effort in the middle of the load and the fulcrum. Pulling or pushing in the middle moves the load on the end.

Levers are used all around us every day. Can you tell which class each one belongs to?

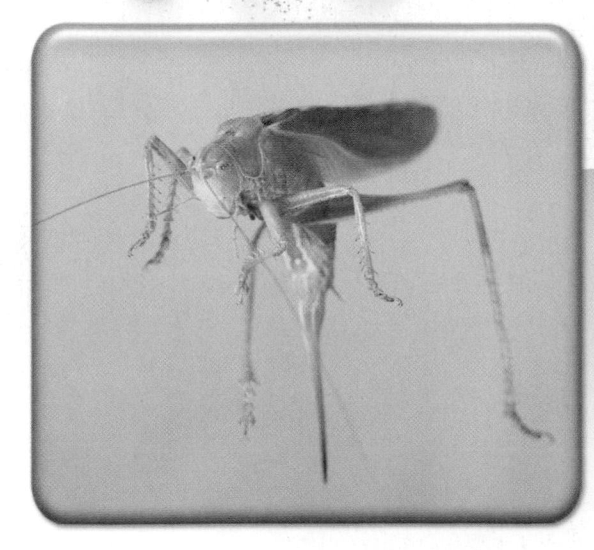

Strange but True

The grasshopper has an exoskeleton with muscles inside it. When one leg muscle shortens, it straightens the lower and upper parts of the leg into one long structure. That powerful lever action thrusts it forward and upward.

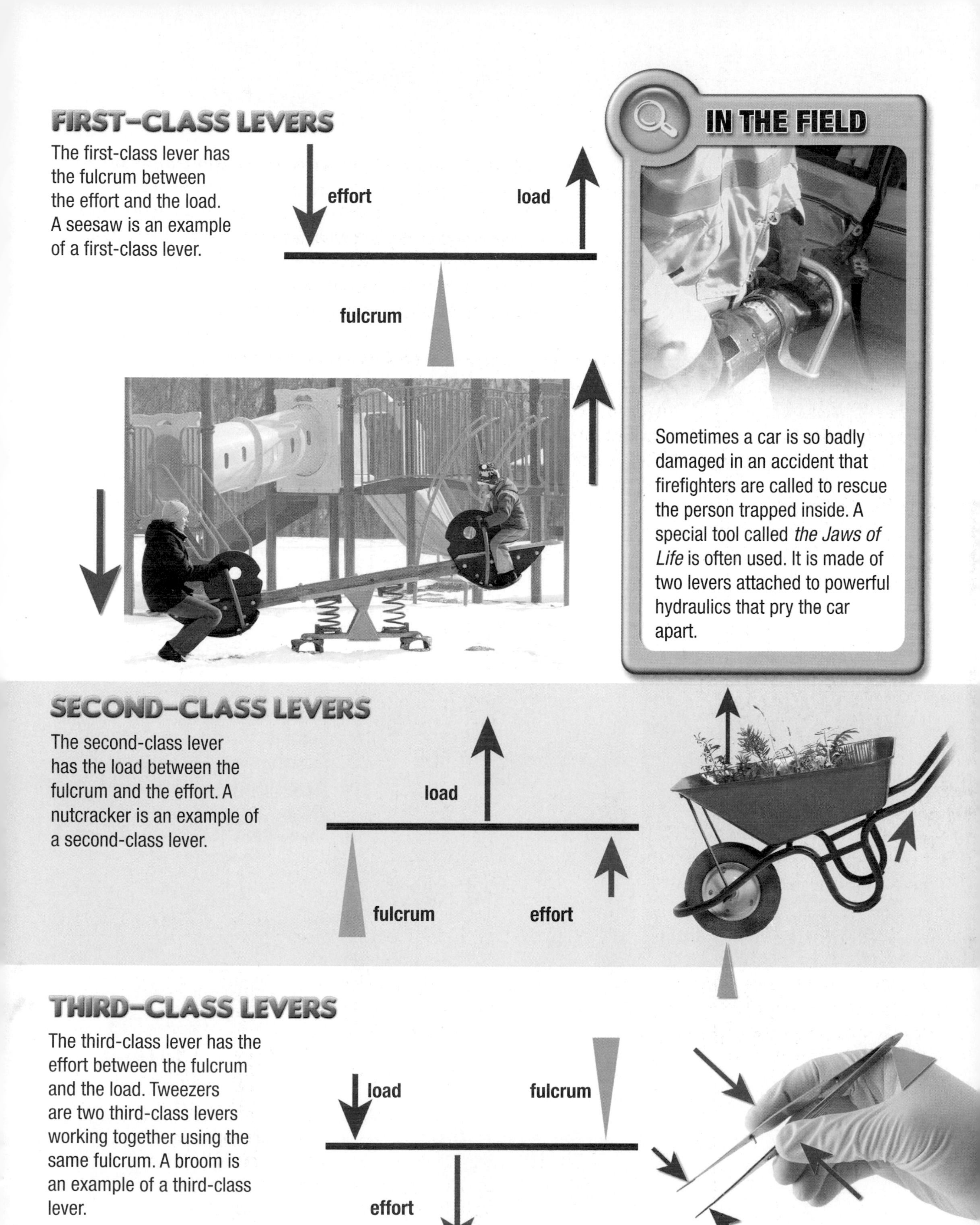

FIRST-CLASS LEVERS

The first-class lever has the fulcrum between the effort and the load. A seesaw is an example of a first-class lever.

IN THE FIELD

Sometimes a car is so badly damaged in an accident that firefighters are called to rescue the person trapped inside. A special tool called *the Jaws of Life* is often used. It is made of two levers attached to powerful hydraulics that pry the car apart.

SECOND-CLASS LEVERS

The second-class lever has the load between the fulcrum and the effort. A nutcracker is an example of a second-class lever.

THIRD-CLASS LEVERS

The third-class lever has the effort between the fulcrum and the load. Tweezers are two third-class levers working together using the same fulcrum. A broom is an example of a third-class lever.

THE ADVANTAGE OF MACHINES

VOCABULARY

grade ('grād) a measure of the steepness of a slope

OUTLOOK

This engraving shows Archimedes trying to lift the world with a lever. It illustrates his famous quote "Give me but one firm spot on which to stand, and I will move the earth." Some people consider this quote to be a metaphor. What if Archimedes wanted to have a great impact on the world? What if the firm spot on which to stand was truth? Logic could then be used like a lever to convince the world to agree on important morals and values. What if everyone believed in the Bible? How would that change the world?

Simple machines have been around for thousands of years. Many ancient civilizations made use of these machines in a variety of ways. The earliest tools found to date are primitive wedges made from stone. The Greeks used ramps in their architecture. Romans built giant siege ramps to get into enemy fortresses, such as the one located at Masada in southern Israel. Many scholars believe that ramps were used in the construction of the ancient Egyptian pyramids.

When a simple machine is used to do work, often the work requires less effort over a greater distance. If this is the case, the machine is providing an advantage to the person using it. Like the lever, the inclined plane can also give an advantage. When the effort on a lever gets farther from the fulcrum, less force is needed to lift the load. Sometimes the length of an inclined plane is increased so that less force is needed to move an object up the ramp.

This is an earthen siege ramp built by the Romans in the first century AD. It was used to give their army access to the mountain fortress of Masada held by a group of Jewish people whose purpose was to resist being ruled by the Roman Empire.

Archaeologists have discovered a variety of artifacts that include the earliest known tools. These primitive stone wedges were made for scraping, cutting, and sawing.

When a longer ramp is used to decrease the amount of effort, the steepness of the slope also decreases. The measure of this steepness is called the grade. A smaller grade means less effort is needed to get to a higher level. Grades are measured as a percentage. A 0% grade means a surface is completely flat. A ramp that rises in height at a 45° angle has a grade of 100%.

The maximum height of the ramp is called *the rise*. The horizontal distance that the ramp covers is called *the run*. To get the grade of the ramp, divide the rise by the run. This results in a fraction. Convert this fraction to a percentage and you have the grade of the slope.

$$\text{grade} = \frac{\text{rise}}{\text{run}}$$

metric

$$\frac{\text{rise}}{\text{run}} = \frac{2.5\text{ cm}}{10\text{ cm}} = \frac{2.5\text{ cm}}{10\text{ cm}} \times \frac{10}{10} = \frac{25}{100}$$

grade = 25%

customary

$$\frac{\text{rise}}{\text{run}} = \frac{1\text{ in.}}{4\text{ in.}} = \frac{1\text{ in.}}{4\text{ in.}} \times \frac{25}{25} = \frac{25}{100}$$

grade = 25%

rise = 2.5 cm (1 in.)

run = 10 cm (4 in.)

Why does an engineer choose not to build a road straight up a mountain?

History Link
The ancient Egyptians lifted giant stones averaging 2,268 kg (2.5 tons) each to build the pyramids. There are several theories surrounding how they did this. Research at least two theories about what tools were used to build these ancient structures. Write a report to summarize your findings.

Technology Link
During the 1800s pioneers in the United States traveled west in covered wagons. Research this topic on the Internet. Identify and describe three tools that were used during travels on the Oregon Trail. Give a report to the class and include pictures of the tools.

Language Arts Link
Write a journal entry of a typical day that highlights the simple machines you used. Include the names of the simple machines, the different tasks performed, and how the effort or distance was changed to complete each task.

It is difficult for wheelchairs to travel up a flight of stairs. Inclined planes are used to access higher levels more easily. Wheelchair ramps in the United States are typically designed with a low grade that is no more than 8.3%. This makes it easier to get up the slope.

CHAPTER 8 ELECTRICITY AND MAGNETISM

KEY IDEAS

- Electricity and magnetism are linked to the atom.
- Electrical energy comes in different forms.
- Magnetism and electrical energy are related.
- There are various ways to produce electricity.

How important is electricity to modern life? There was a time when traveling any distance required a person to either walk or ride in a carriage pulled by an animal like a horse or an ox. Today, electrical energy powers vehicles, such as cars and even small airplanes. Rather than drying clothes on a line, clothing is dried faster and more thoroughly in electric dryers. Electrical energy has changed people's lives, often for the better. How does it work? How is magnetism related to electrical energy? Are there different ways to produce electricity that will cause less harm to the environment? These and other questions require an investigation of electrical energy and magnetism, their relationship, and how they are used.

Strange but True

How does this object produce electricity?

Windpower

OUTLOOK

When God wanted to demonstrate His sovereignty to Job, He asked him a series of questions. "What is the way to the place where the lightning is dispersed, or the place where the east winds are scattered over the earth?" (Job 38:24) Job had to confess that he did not know. Only God knows everything and is mighty enough to rule over every part of creation, from electrical storms to blustery winds. If God can control something as powerful as a bolt of lightning, is there anything He cannot do?

STATIC ELECTRICITY

VOCABULARY

charge ('chärj) the property of a particle that causes it to attract or repel other particles

static electricity ('sta·tik i-ˌlek-'tri-sə-tē) the buildup of nonflowing electric charge

static discharge ('sta·tik 'dis·chärj) a sudden flow of static electricity from one object to another

Electricity involves charges. A charge is a property of an atomic particle that causes it to attract or repel other particles. Opposite charges attract one another. Particles with the same charge repel one another. This is the basic principle behind electrical energy and magnetism.

At the center of an atom is its nucleus, which is made up of protons and neutrons. Protons are positively-charged particles. Neutrons are particles with no charge. Electrons are negatively-charged particles. While a proton is bound securely in the atom's nucleus, an electron orbits the nucleus in a spherical region. An electron can sometimes leave its atom. When an atom has the same number of protons and electrons, the opposite charges cancel each other out so that the overall charge of the atom is neutral.

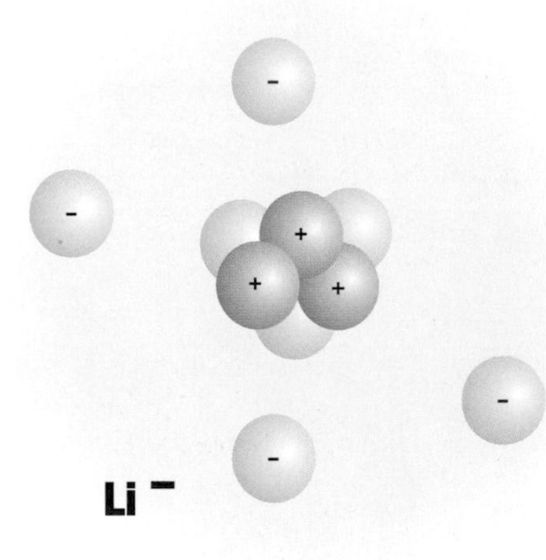

Li

Li^{+}

When a neutral lithium atom gains or loses an electron, it is called *a lithium ion*. If the atom gains an electron, it will have an overall negative charge. If it loses an electron it will have an overall positive charge. Lithium is used to make certain kinds of batteries. As a medicine, its ions also help to moderate mood swings in people with mental disorders.

Most objects have no overall electric charge because the majority of the atoms that compose them are neutral. Sometimes an object can become charged by gaining or losing electrons. For example, if you rub an inflated balloon against your sweater, some of the electrons from the sweater will transfer to the balloon. This will leave the sweater with more protons than electrons, giving it an overall positive charge. Since the balloon has now gained electrons, it will have an overall negative charge. This buildup of electric charges is called static electricity. If you hold the balloon near your hair, its negative charge will attract the protons in your hair. This is why your hair sticks to the balloon.

Static electricity is considered one form of electricity. *Static* means *at rest*. When two oppositely charged objects come into contact, electrons will transfer between them until both objects are neutral. This transfer is called static discharge. It is what you experience when you walk across a carpet, touch a metal object, and feel a slight electric shock.

QUICK FACT

Lightning is static discharge. The air in storm clouds swirls rapidly and causes water droplets to gain and lose electrons. This makes the droplets electrically charged. If a storm cloud develops an overall negative charge, it can cause the ground beneath it to become positively charged by repelling some of its electrons. Lightning transfers billions of trillions of electrons between the cloud and the ground.

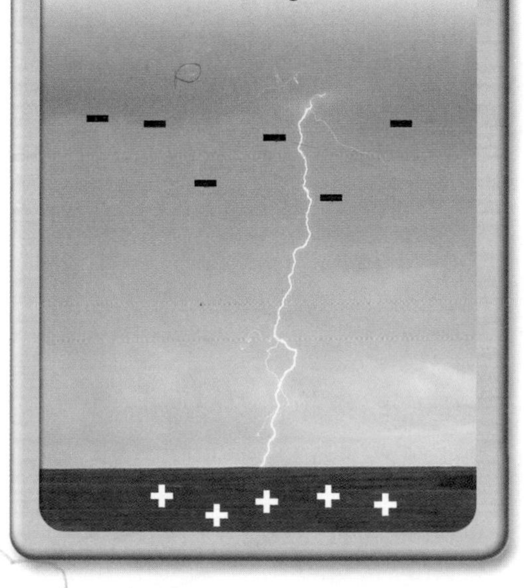

Electrons transfer to other atoms and objects because they are not held tightly to the nucleus as protons and neutrons are.

CURRENT ELECTRICITY

VOCABULARY

current electricity ('kûr·ənt i·ˌlek·'tri·sə·tē) the continuous flow of electric charge through a material

circuit ('sûr·kət) a complete path through which an electric current can flow

voltage ('vōl·tij) the power of an electric current, measured in volts

Static discharge is a sudden, temporary flow of electric charge. In contrast, current electricity is a continuous flow of electric charge. It is another form of electricity and can flow through certain types of material. Many metals, such as copper and aluminum, contain electrons that are only loosely bound to their atoms. This makes these materials good conductors because the loose electrons easily form an electric current. Materials like plastic, rubber, and glass are insulators. They resist the flow of current electricity because their electrons are tightly bound to their atoms.

Just because a material is a conductor does not mean that an electric current is flowing through it. For current electricity to flow continuously, it must follow an unbroken circuit. If it is broken in any place, no electric current can move through it.

Why are these wires made of two different materials?

If this track had a gap anywhere on it, the roller coaster could not run. Similarly, current electricity must have an unbroken circuit in order to flow.

A copper lightning rod like this one is an excellent conductor. When lighting strikes, it conducts the static discharge safely into the ground. This protects the rest of the structure from damage.

Not only does current electricity require a circuit in order to flow, it also must have energy. The difference in potential energy between two ends of a circuit is what allows an electric current to flow. This is called voltage. A source of voltage like a battery has two ends with opposite charges. Electrons build up on the negatively charged end. When the two ends are connected by a circuit, the electrons at the negatively charged end flow through the circuit toward the positively charged end.

Imagine a downhill skier at the top of the slope. What creates the energy to ski downhill? When a skier rides the ski lift to move up the hill, potential energy increases because of the force of gravity. This force provides the potential to ski down the hill. The higher up the hill the skier goes, the more potential energy there is. The difference in the potential energy between the top of the hill and the bottom of the hill creates the energy that makes skiing so exciting.

The difference in potential energy between the two ends of this battery is what allows an electric current to flow between them when they are connected by a circuit.

IN THE FIELD

A lineman is an electrician who helps maintain high power lines. A hydraulic lift called *a cherry picker* lifts the lineman up to where the line needs repairing. This can be a dangerous job because of the high amount of current flowing through the wires. These workers are specially trained to be able to safely repair power lines that have high voltages and currents.

A chairlift transports skiers to the top of the ski hill. The potential energy they have because of their position allows them to ski quickly down the hill.

CIRCUIT TYPES

VOCABULARY

resistance (ri·ˈzis·təns) a measure of how difficult it is for an electric current to flow

series circuit (ˈsîr·ēz ˈsûr·kət) a circuit with a single path for an electric current

parallel circuit (ˈpâr·ə·lel ˈsûr·kət) a circuit with more than one path for an electric current

The amount of current in a circuit is affected by several factors. Increasing the voltage will increase the amount of current. Decreasing the amount of voltage will decrease the amount of current. Resistance is another factor that determines the amount of current. As resistance increases, the amount of current decreases. As resistance decreases, the amount of current increases.

The material a circuit is made of is a factor that affects resistance. Conductors offer little resistance. If a circuit is made out of a good conductor, the current will be greater than if it were made out of a poor conductor. If a circuit is long, the current will have to travel farther. This increases resistance. If the circuit is thin, it will have more resistance than if it is thick. A thin circuit has less area for the electric current to flow through.

What would increase this current of water?

This bird can sit on a bare wire without fear of electric shock. The bird's body offers more resistance than the wire, so the current flows through the wire instead of the bird.

Fire fighters use hoses that are wider than a normal garden hose. This decreases the resistance and allows more water to flow through the hose to fight the fire.

There are different ways to arrange a circuit. A series circuit has only one path for an electric current to flow through. Any break in the circuit will stop movement of the current. If you connect two lightbulbs to a battery using a series circuit, the same electrons will pass through each lightbulb. The second bulb increases the resistance in the circuit. This is because the current must travel farther and must go through one bulb before reaching the other. Additional bulbs on the same circuit will not burn as brightly as a single bulb. If one of the lightbulbs is removed or burned out, it causes a break in the circuit. The second bulb will not glow.

In a parallel circuit the electric current has more than one path. If two lightbulbs are connected on a parallel circuit, removing one will not affect the other. The current will simply flow through the other pathway. Also, adding bulbs does not increase the amount of resistance as much as it does in a series circuit. This is because both bulbs are connected directly to the power source rather than having a source of resistance between them and the source of power.

Does turning off your lamp also turn off your television?

QUICK FACT

In the past the lights for Christmas trees were often constructed on a series circuit. What do you think happened to the string if one bulb burned out? People spent a great deal of time searching through dozens of little bulbs in order to replace one burned-out bulb so that the rest would work. Designing strands of Christmas lights using parallel circuits helped to reduce this particular holiday frustration.

Series Circuit

A B

If Lightbulb B is removed, it causes a break in the circuit. Lightbulb A will go out too.

A B

Parallel Circuit

A B

If Lightbulb A is removed from the circuit, the current will flow through the other branch. Lightbulb B will still remain lit.

A B

ELECTROMAGNETISM

VOCABULARY

ferromagnetic
(ˌfâr·ō·mag·ˈne·tik) relating to matter with strong magnetic properties

electromagnet
(i·ˌlek·trō·ˈmag·nət) a magnet that produces a magnetic field by means of an electric current

Magnetism is an invisible force that attracts objects made of iron or related metals. Electromagnetism is magnetism produced by an electric current. Both current electricity and magnetism exist because of the movement of electrons in atoms. When electrons move from one atom to another, it creates an imbalance or flow. This movement causes an electrical current. Electrons in certain atoms produce magnetic fields. In most atoms, electrons are paired. In some types of matter, atoms have unpaired electrons, which causes the overall atoms to have magnetic properties.

Iron, nickel, and cobalt are elements that have magnetic characteristics. If an object is strongly magnetic, it is called ferromagnetic. Just as positive and negative charges attract each other, the north end of a magnet and the south end of another magnet attract each other. Just as two positive or two negative charges repel each other, the north ends of two magnets or the south ends of two magnets also repel each other.

Is this magnet growing fur?

Magnetite is a ferromagnetic mineral. This natural magnet is a compound of iron and oxygen.

Magnetic resonance imaging (MRI) is a medical procedure used to identify tumors, bleeding, injury, and even infection. A patient is placed inside the machine, which uses magnetic fields and radio waves to show internal organs and organ systems.

Any moving electric charge produces a magnetic field. Electrons flowing through a wire create a magnetic field around the wire. When the wire is looped around something magnetic, like an iron nail, the magnetic field can be very strong. The device is called an electromagnet. It will have a north pole and a south pole that can be detected by a compass. As long as electric current is flowing, the electromagnet will generate a magnetic field. When the current stops, the magnetic field no longer exists.

Electromagnets are used to store information on computer hard drives and credit cards. Doorbells, motors, loudspeakers, and MRI machines all use electromagnets. People who work in some scrap yards use giant electromagnets to lift the cars and move them from one place to another.

A crane with an electromagnet is used to pick up scrap metal. When the crane operator turns off the electric current, the magnetism stops and the scrap metal is released.

Hard drives, card scanners, bark collars, and earbuds all use electromagnetism.

PRODUCING ELECTRICITY

VOCABULARY

generator ('je·nə·rā·tər) a machine that produces electricity from moving parts

Electrical energy can be transformed into other kinds of energy. For example, an electric oven uses electrical energy to produce the thermal energy needed to cook food. A flashlight uses a battery to produce electrical energy to create light. A battery-powered toy can turn electrical energy into the kinetic energy of motion as it moves, or into sound and light as it flashes and makes noise. Electrical energy runs machines in factories and hospitals and powers heaters that keep people warm in cold weather.

Various forms of energy can be used to produce electrical energy. A power plant uses huge generators to make electricity. The plant heats water to create steam. The movement of the steam pushes a set of wire coils through magnetic fields. The magnetic fields force electrons to flow through the wires. The current electricity produced flows through power lines that are connected to people's homes and businesses by smaller wires.

Strange but True

Wind turbines are generators. They convert the kinetic energy of wind into electrical energy.

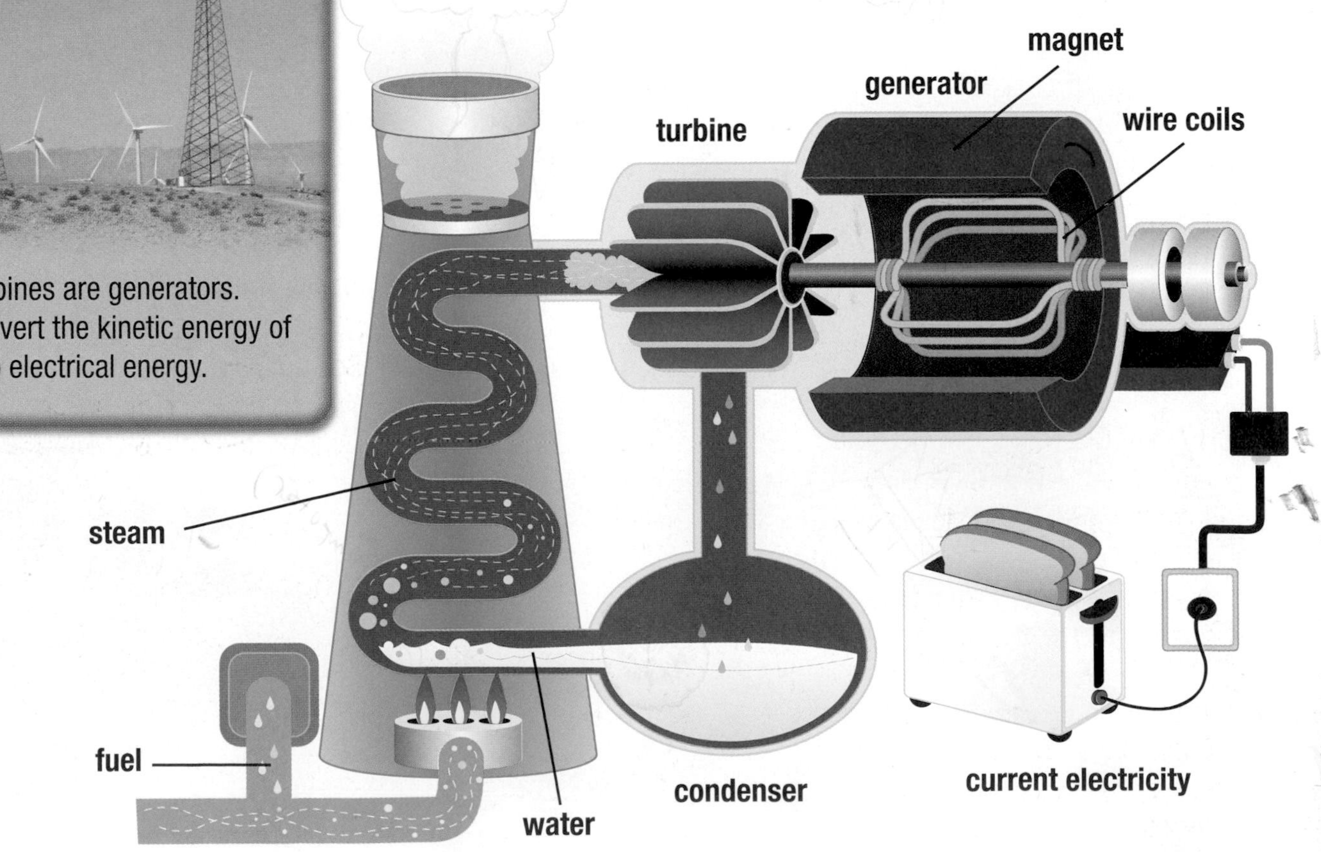

Usually power plants burn coal, oil, or natural gas in order to boil the water that runs their generators. A nuclear power plant uses nuclear energy to boil the water. Unfortunately, these methods have some disadvantages. The burning of fuels creates pollution and uses up these sources of energy. Both types of plants must cool the boiled water before they can reuse it or return it to the environment. Nuclear energy must be produced in a highly controlled setting. However, it is less costly to produce than energy generated by the burning of fuels.

Some power plants are choosing other methods to make electricity. Wind farms, for example, use the kinetic energy of strong, consistent breezes to turn wind turbines. Hydroelectric plants harness the kinetic energy of falling water. Both of these kinds of plants transform kinetic energy into electrical energy. Some people use solar energy to power their homes and businesses.

Why should a power plant cool water before returning it to a river?

LINKS

Bible Link
Look in the Old Testament for passages that describe lightning. Pretend that you are a person who lived in Bible times. Write a story in the first person that describes how you feel about lightning and what its power demonstrates about God's majesty.

Technology Link
Research how electromagnets are involved in technology. Create a poster that shows tools and equipment that use electromagnetism. Provide descriptive captions for each illustration.

History Link
Research Thomas Edison's electrical inventions. Choose one and prepare a computer presentation or illustrate a comic book explaining why it was invented and how it works.

Solar panels collect energy from the sun. This energy can be transformed into electrical energy.

These are hydroelectric generators. Rather than boiling water to make energy, hydroelectric plants use the kinetic energy of flowing water.

LIFE SCIENCE: CYCLES

UNIT 2

PHYSICAL SCIENCE: TRANSFORMATIONS

EARTH AND SPACE SCIENCE: PREDICTABILITY

– the study of Earth systems and systems in Space

HUMAN BODY: BALANCE

CHAPTER 9

EARTH'S PROCESSES

Minerals are solid, inorganic, naturally occurring substances found on Earth. They have a definite chemical composition and crystal structure. By themselves, minerals have many uses. For instance, they make up beautiful gemstones in jewelry such as diamonds, rubies, emeralds, sapphires, and quartz. From chalk to graphite, minerals are at work in your classroom. Halite, the mineral name for table salt, or sodium chloride, flavors and preserves food. Talc can be ground up and used as baby powder. Many different minerals in the foods you eat help your body to work properly.

KEY IDEAS

- Rocks change in a predictable pattern known as *the rock cycle*.
- The plate tectonics theory describes how the earth's crust is constantly being recycled.
- The ocean floor contains a variety of surface features.
- Density, salinity, and temperature affect ocean currents and global climate patterns.

The earth is always changing—on land and sea and even beneath the ocean floor.

Tourmaline forms when magma cools deep under Earth's surface.

These gypsum crystals are from the Kara-Kum Desert in Turkmenistan in Central Asia. Gypsum crystals can form as mineral solutions on the surface of the earth evaporate.

Metamorphic rock forms when large amounts of heat and pressure are applied to other rock types. The amount of change that these rocks undergo depends on the amount of pressure they experience. Examples of metamorphic rocks are marble, slate, and gneiss.

Rocks transform from one type to another, depending on the conditions they experience. Understanding the rock cycle helps scientists predict how rocks will change. For instance, igneous rock forms when one of the other types melts, cools, and hardens. For sedimentary rock, wind and water must erode small pieces of other rock particles, which are then layered, compacted, and cemented together. Before igneous or sedimentary rock can change into metamorphic rock, tremendous amounts of heat and pressure must be added. If more heat and pressure are added to rock that is already metamorphic, it stays metamorphic. It simply exhibits a higher degree of change, sometimes showing bands of mineral deposits.

OUTLOOK

Change is a part of life, except for God who does not change (Malachi 3:6). Caterpillars go through metamorphosis to become butterflies. Even inorganic material like rock experiences alteration that results from heat, pressure, and weathering. Being flexible and able to cope with change is an important character trait for people to lead successful lives.

Metamorphic rocks go through varying amounts of change, depending on how much heat and pressure they experience. This slate was once sedimentary shale, formed from clay, silt, or mud.

...s like this
In sedimenta...e particles that
red sandst... and harden into
come t...en clearly visible.
rock...

RECYCLING THE CRUST

VOCABULARY

lithosphere (ˈlith·ə·sfîr) the crust and solid, upper portion of the mantle

The earth's crust is a solid, rocky layer, made up of igneous, sedimentary, and metamorphic rocks. Below the crust are three different layers—the mantle, outer core, and inner core. Geologists call the crust and upper mantle the lithosphere. The plate tectonics theory states that the lithosphere is divided into plates that float on the flowing lower mantle. These lithospheric plates interact with one another at their edges. Interactions between plates can cause new crust to be formed or existing crust to change form and be recycled. Unfortunately, plate movements can sometimes cause earthquakes, volcanoes, and tsunamis.

The plate tectonics theory was partly shaped by the continental drift theory, which was proposed by German meteorologist Alfred Wegener. He thought that the continents looked like puzzle pieces that were once connected. This idea is known as *Pangaea*. The idea of one supercontinent claims that the continents split apart and drifted over the ocean's crust. There is debate about whether or not Pangaea existed, and if so, how long it took to separate.

Strange but True

Four tectonic plates form boundaries under the Atlantic Ocean. Scientists have determined that this seafloor spreads outward about 2 cm (1 in.) per year, which is slightly less than your fingernails grow from one year to the next.

Earthquakes can have devastating results. In 2011 this destruction was caused by an earthquake just off the coast of Japan. Earthquakes can trigger volcanoes when magma comes up through big cracks in the earth's crust.

Three types of boundaries exist as a result of plate movement. Convergent boundaries occur when two plates collide. This is one way that continental and oceanic mountains are built as two plates meet and push up the crust. Colliding plates can also cause ocean trenches to form when one plate is more dense than the other. Old crust changes form and is often recycled as it plunges into Earth's hot mantle to melt, eventually forming igneous rock.

A divergent boundary occurs when two plates pull apart. Magma fills in the gap, forming new crust as it cools. This movement can explain rifts on continents, ridges in oceans, and volcanoes in both.

Transform boundaries relieve some tension on lithospheric plates as they slide past one another. This movement usually only causes small cracks. However, sometimes faults and earthquakes occur along transform boundaries.

QUICK FACT

This area of Indonesia was devastated by the Indian Ocean tsunami in December 2004. It was caused by an earthquake with a magnitude of 9.1 on the Richter scale. This was the largest tsunami in recorded history. Approximately 225,000 people died as a result of it.

What type of boundary would most likely have formed this mountain?

Station Owners
- NDBC DART
- Australia
- Chile
- Ecuador
- Thailand

These diamonds represent the location of DARTs—devices to forecast tsunamis. Notice how their locations compare to the boundaries of lithospheric plates. The Ring of Fire is a zone at the edges of the Pacific Plate where many volcanoes erupt and earthquakes occur.

DART (Deep-Ocean Assessment and Reporting of Tsunamis) buoys are positioned in areas where tsunamis are a significant threat to land areas. These tools help scientists accurately predict when and where a tsunami may strike land. The goal is to provide people with advanced warning.

THE OCEAN FLOOR

VOCABULARY

abyssal plain (ə·ˈbi·səl ˈplān) a large, nearly flat region on the ocean floor

seamount (ˈsē·mount) an undersea volcanic mountain found on the ocean floor

subduction (səb·ˈduk·shən) the process by which one lithospheric plate is forced under another

The invention of the echo sounder in the early 1900s allowed scientists to get a more accurate and complete view of the ocean floor. An echo sounder is an instrument placed on the bottom of a ship that sends sound waves to the ocean floor. The sound waves bounce off the seafloor and return to the device. Scientists note the time it takes for the sound waves to make the round trip. They use this data to calculate the depth of the ocean floor. In the 1950s and 1960s, scientists discovered some amazing oceanic features using this method. They found underwater mountain ranges that were longer and higher than those on the continents. They also discovered volcanoes, ridges, and deep ocean trenches. Most importantly, scientists learned about movement under the ocean's surface that causes the ocean floor to change form.

How does this animal use sonar?

Sonar stands for Sound Navigation And Ranging. It is a device used to locate objects. Scientists utilize it today to map the ocean floor. Fishermen locate certain marine life with it. Submarine crews use sonar to guide them through deep water.

Mauna Kea is an inactive volcano in Hawaii. It is the tallest mountain on Earth. Although only about 4,206 m (about 13,800 ft) of it rises above sea level, it actually stands about 10,211 m (33,500 ft) tall when measured from the ocean floor!

The ocean is divided into three basic areas—continental margins, deep ocean basins, and ocean ridges. The continental margins include the shelf and slope, which are underwater extensions of the continents. The continental shelf is the land under water that borders the continents. The shelf extends on average 65 km (40 mi) from the continents toward the continental slope.

The deep ocean basins contain several features. The abyssal plains are large, nearly flat sections that make up most of the ocean floor. Seamounts are underwater volcanic mountains that form near divergent plate boundaries or hot spots. Hot spots are areas of intense heat in Earth's mantle that cause magma to rise onto the surface of the inner areas of lithospheric plates.

Trenches are underwater valleys where old ocean floor changes form. As two plates converge, the denser plate is forced under the other in a process called subduction. The subducted plate melts as it gets closer to Earth's mantle. Seafloor spreading occurs when plates diverge, forming ocean ridges as magma fills in and builds up in the open spaces.

IN THE FIELD

Members of HMS *Challenger* performed a study of the ocean floor's surface features from 1872–1876. They determined the depth of the ocean by repeatedly lowering a weighted line. This was very hard work and took a long time to accomplish. The crew also collected organisms that biologists would later analyze. For the next 75 years, much of what was known about the ocean was a result of this four-year journey.

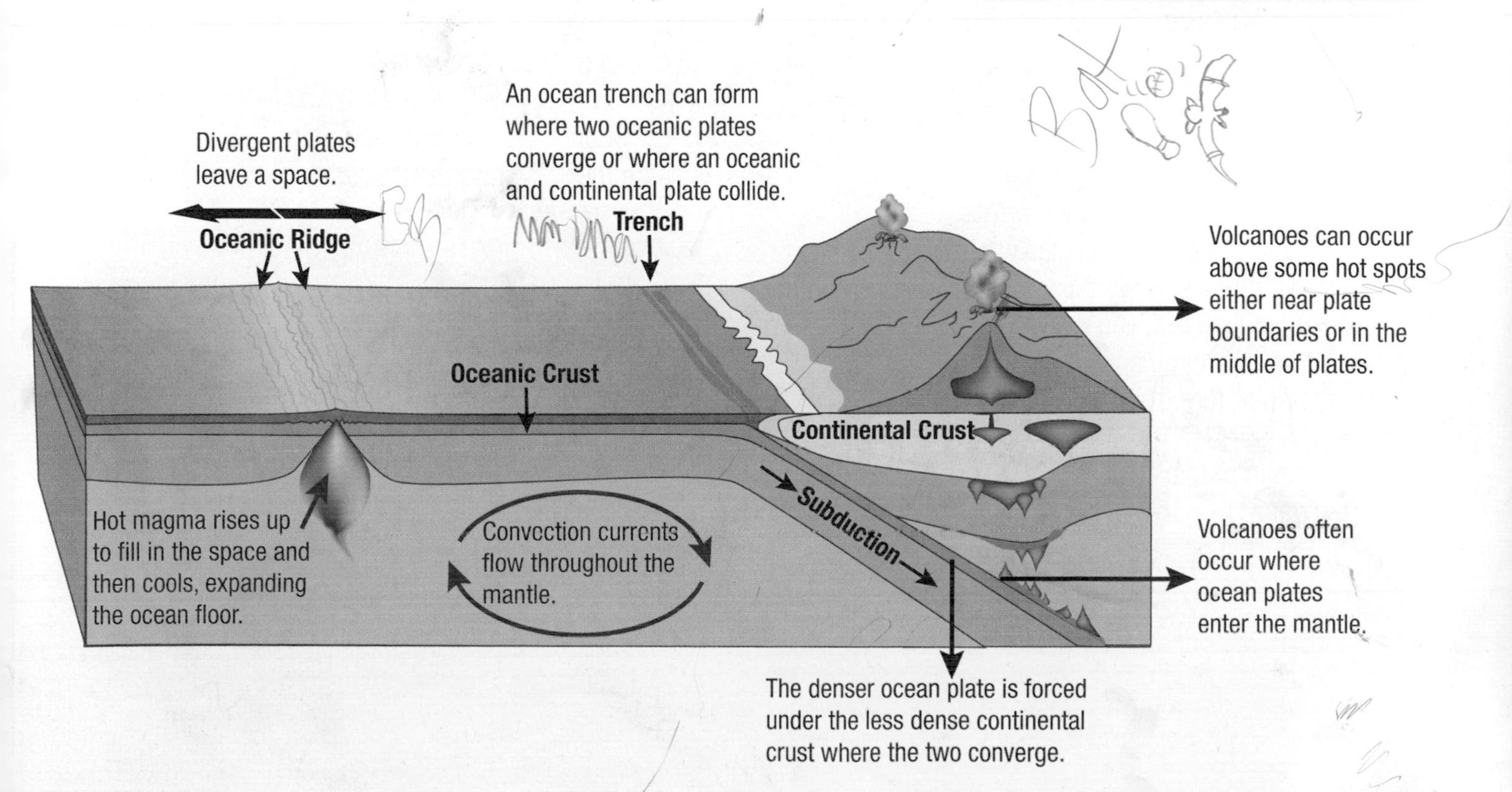

OCEAN CURRENTS

VOCABULARY

deep ocean current ('dēp 'ō·shən 'kûr·ənt) a stream of water that flows deep beneath the ocean's surface

Coriolis effect (kôr·ē·'ō·ləs i·'fekt) the way Earth's rotation affects wind and ocean currents

El Niño ('el·'nē·nyō) a climate event in the Pacific Ocean caused by wind shifts, resulting in unusually warm water temperatures

upwelling (əp·'we·ling) the movement of cold, deep ocean water to the ocean's surface

Air and ocean currents distribute heat throughout the earth. Surface ocean currents are driven mostly by temperature and wind. Beneath the ocean's surface, deep ocean currents move because of water's gravity and density. The ocean's density is affected by its temperature and the amount of salt in it. The colder the temperature, the more dense the water becomes. Seawater's density also increases as its saltiness, or salinity, increases.

Ocean water exists in layers. The coldest and saltiest water moves to the bottom. Warmer, less salty layers form above. Currents develop when layers move faster or slower than their surrounding layers. Warm surface currents flow toward the earth's poles where some of the water freezes. This freezing causes higher salinity in the remaining water. This denser water sinks and flows back toward the equator.

Ocean currents move because of the earth's rotation. The ocean currents curve clockwise in the Northern Hemisphere and counterclockwise in the Southern Hemisphere. The effect of the earth's rotation on wind and ocean currents is called the Coriolis effect.

Ocean currents carry water across long distances. NASA uses satellites to take pictures of ocean currents. The Gulf Stream is easily visible as the warmest water in the image. It is shown in red. The warm waters, in orange and red, range in temperature from 22°C–32°C (72°F–90°F). The coldest waters are shown in purple and blue and range between 0°C–15°C (32°F–59°F).

When ocean currents do not flow according to their normal patterns, irregular weather can happen. For example, El Niño is an unusual event that occurs every two to seven years. It lasts about 18 months and results from changes in strong Pacific trade winds near the equator. Trade winds are winds that blow steadily in one direction for a long period of time. The winds cause large amounts of warm surface water to move toward the South American coast. This event causes shifts in weather patterns around the world that can include warmer winters, heavy rains, floods, droughts, and even tornadoes.

Sometimes winds cause water deep in the ocean to mix with surface waters. This creates an upwelling, which brings to the surface many nutrients that fish eat. Without this motion the surface waters of the open ocean would not have enough of these nutrients. El Niño prevents upwelling from occurring, which causes fish to die or to relocate.

QUICK FACT

Jason-1 was a satellite that orbited Earth from 2001–2013. From far above the earth, it collected data on ocean temperatures, surface heights, and currents. This spacecraft helped to track El Niño events. In 2008 a team of US and European scientists launched Jason-2 into orbit to continue collecting data.

What kinds of conditions does a fish need to survive?

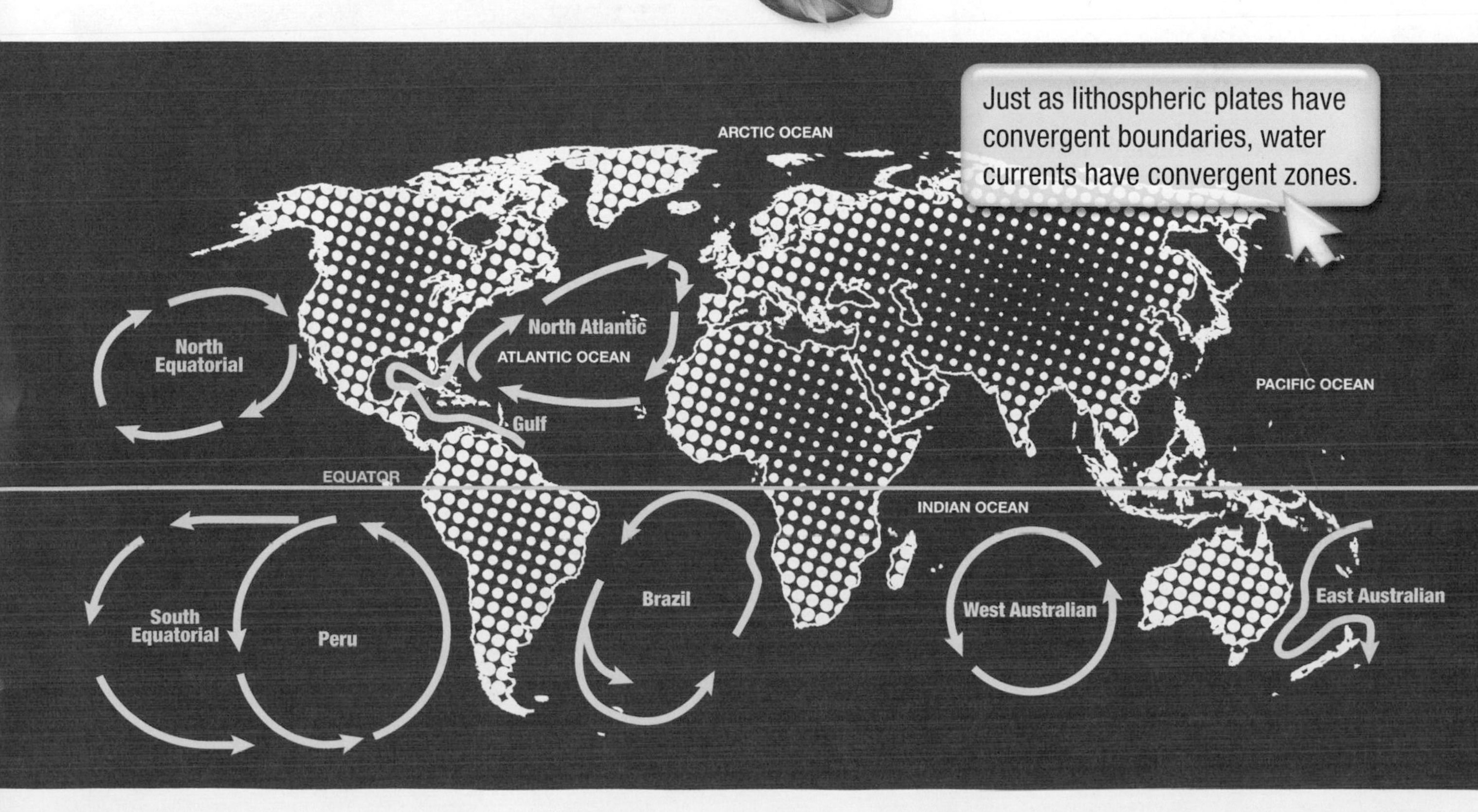

Just as lithospheric plates have convergent boundaries, water currents have convergent zones.

MAGNETIC REVERSALS

VOCABULARY

polarity (pō·'lâr·ə·tē) the quality of having two opposite poles—one positive and one negative

IN THE FIELD

The study of the changes of the earth's magnetic field is called *paleomagnetism*. Paleomagnetists analyze samples taken from the ocean floor to observe the spreading of the sea floor. By taking core samples and looking at the layers of rocks, geoscientists propose hypotheses about the earth's formation and organisms that lived in the past.

Earth's magnetic North and South Poles are not the same as its geographic poles. The geographic North and South Poles remain stationary, but the magnetic poles move. Some geologists believe that the magnetic poles have actually reversed their positions several times in the distant past.

Hikers often use compasses to determine which direction is north. A compass works because convection currents inside the earth make the planet like a giant bar magnet. A hiker in the woods might use some organisms as natural compasses. For example, in certain climates moss grows thickest on the north side of trees. Also, the position of stars or the sun can help people and animals navigate.

Even rocks serve as natural compasses. A mineral called *magnetite* contains molecules that always point toward magnetic north. When scientists find magnetite in molten igneous rock, they observe that its particles move to align themselves toward the north. Once the magnetite cools, its molecules stay in place.

How do these birds know how to fly in the right direction?

The compass plant grows on the prairies of the United States. Its leaves are aligned in a north-south direction. This vertical arrangement allows both surfaces of the compass plant's leaves to be involved in photosynthesis.

By looking at samples of the crust in the Atlantic Ocean, scientists made some unique discoveries. They found alternating stripes of igneous rock. Every other stripe had most of its magnetite molecules facing magnetic north. This is known as normal polarity. The remaining stripes contained magnetite molecules where the overall magnetism was in the opposite direction. Scientists call this *reverse polarity*.

The seafloor spreading theory states that the alternating bands of rock resulted from the formation of new crust at the center of the mid-ocean ridge. The Mid-Atlantic Ridge occurs at a divergent boundary in the Atlantic Ocean. Magma rises up through the gaps of the lithospheric plates and hardens. Over time the seafloor spreads out with new crust appearing along the boundary.

LINKS

Technology Link
During this last century, scientists have advanced in their understanding of the ocean floor. After obtaining permission from a parent or guardian, use the Internet to look up recent scientific underwater drilling ventures. In three paragraphs briefly summarize what you discover about the ships, people, and organizations that are working together to learn more about this mysterious, watery frontier.

Language Arts Link
Imagine that you are a rock in Earth's rock cycle. You have been asked to journal your transformation. Write a newspaper article about your life's journey. Use at least three vocabulary words from this chapter in your piece. Include an illustration of the rock or make a model of it.

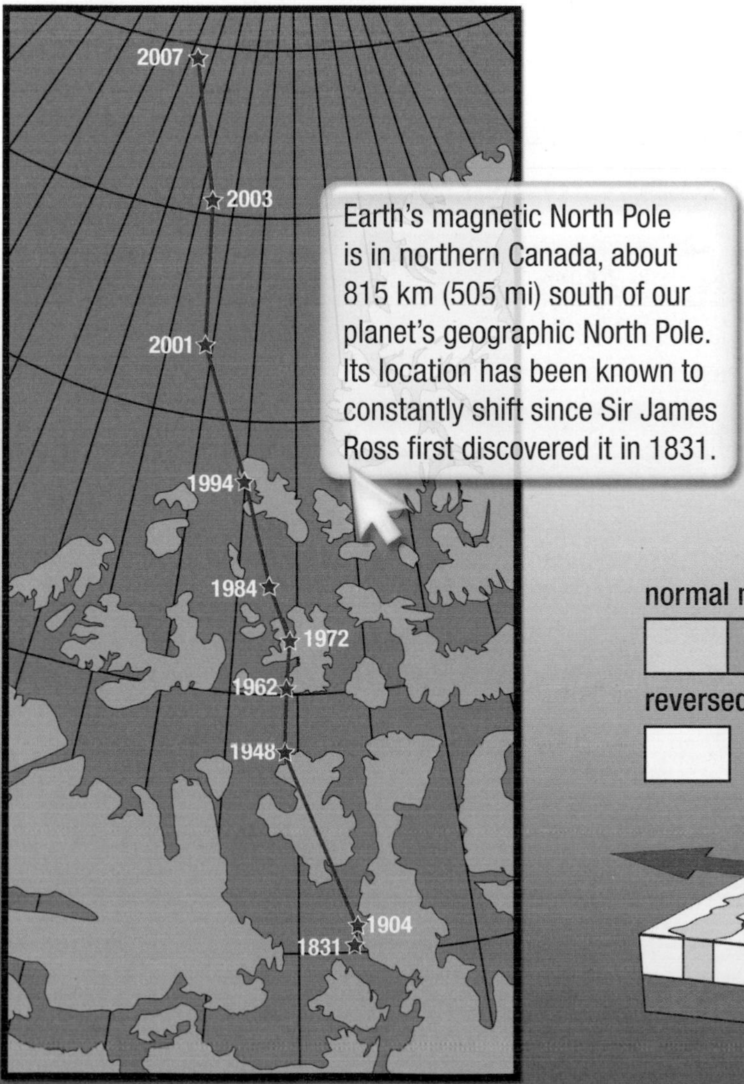

Earth's magnetic North Pole is in northern Canada, about 815 km (505 mi) south of our planet's geographic North Pole. Its location has been known to constantly shift since Sir James Ross first discovered it in 1831.

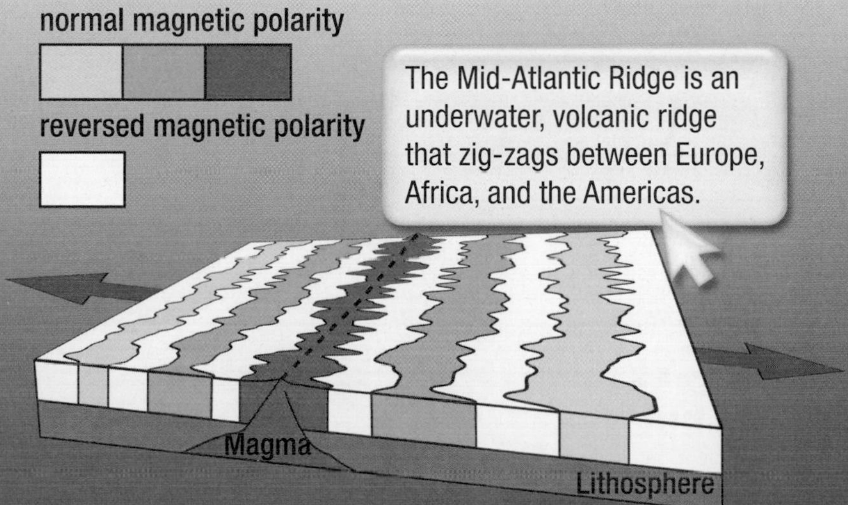

2007
2003
2001
1994
1984
1972
1962
1948
1904
1831

normal magnetic polarity
reversed magnetic polarity

The Mid-Atlantic Ridge is an underwater, volcanic ridge that zig-zags between Europe, Africa, and the Americas.

Magma
Lithosphere

CHAPTER 10 NATURAL RESOURCES

When God created Earth, He included the materials people need to survive. Air, food, and water, as well as petroleum, trees, minerals, wind, and animals are some examples of the natural resources available. People change many of these materials into products that make life more comfortable. Various natural resources are used to generate energy. These are called *energy resources*. Since all these materials that God has provided are necessary for life, it is important to be good stewards of them.

VOCABULARY

natural resource (ˈna·chə·rəl ˈrē·sôrs) a material found in nature that is useful to humans

KEY IDEAS

- Natural resources are either renewable or nonrenewable.
- Some natural resources are used to generate energy.
- Minerals have many uses and are a nonrenewable resource.
- Reduce, reuse, and recycle are three ways to conserve natural resources.

Wind is a resource that is harnessed to generate energy.

Minerals are mined for use in jewelry, food, transportation vehicles, building materials, and electronics.

Strange but True

Would you eat this?

QUICK FACT

Humans use minerals in almost everything they do. According to the Mineral Information Institute, almost 21,800 kg (48,000 lb) of minerals must be supplied for each person in the United States every year. Over the course of a lifetime, it is estimated that an individual will use about 1.7 million kg (3.7 million lb) of minerals, metals, and fuel.

MINERALS, METALS, AND FUELS
USED PER PERSON IN A LIFETIME

- 311,044 L (82,169 gal) petroleum
- 780,179 kg (1.72 million lb) stone, sand, gravel
- 14,812 kg (32,654 lb) iron ore
- 262,610 kg (578,956 lb) coal
- 161,689 m^3 (5.71 million ft^3) natural gas
- 14,474 kg (31, 909 lb) salt
- 8,367 kg (18,447 lb) phosphate rock

Soil is considered a natural resource because it is a material found in nature that people use for growing crops, gardening, landscaping, brick-making, and managing waste.

Paper products and some building materials are made from trees. A variety of trees produce fruit and others are burned as fuel.

Water is essential to life and provides energy.

VOCABULARY

renewable (ri·ˈnoo·ə·bəl) capable of being replenished or replaced

desalination (dē·ˌsa·lə·ˈnā·shən) the process of removing salt from seawater to obtain freshwater

nonrenewable (ˌnän·ri·ˈnoo·ə·bəl) not capable of being replenished or replaced within a sufficient period

Natural resources are often classified into two types according to whether or not they can be renewed. Renewable resources can be replaced at the same rate at which they are consumed. Solar energy, trees, water, soil, and animals are examples of renewable resources. Although these resources are renewable, it is possible to use them faster than they can be replaced.

Water is a very precious natural resource. It is renewed when rain or snow fills lakes, rivers, and reservoirs. In some areas of the world, people desalinate ocean water because there is a limited supply of freshwater. Desalination is the process by which salt is removed from seawater. The freshwater is then collected for various uses. Desalination typically requires large amounts of energy as well as expensive equipment.

OUTLOOK

Trees are an amazing gift from God. They provide shade and homes for many different kinds of animals. They remove pollutants from the air, reduce smog, and diminish noise pollution by absorbing sound waves. They also replenish oxygen, give off moisture, and lower the air temperature through transpiration. Trees play an intricate role on Earth and are valuable natural resources.

Desalination is one of the earliest forms of water treatment. Today, this process is still used on ships and on land to convert seawater into fresh drinking water.

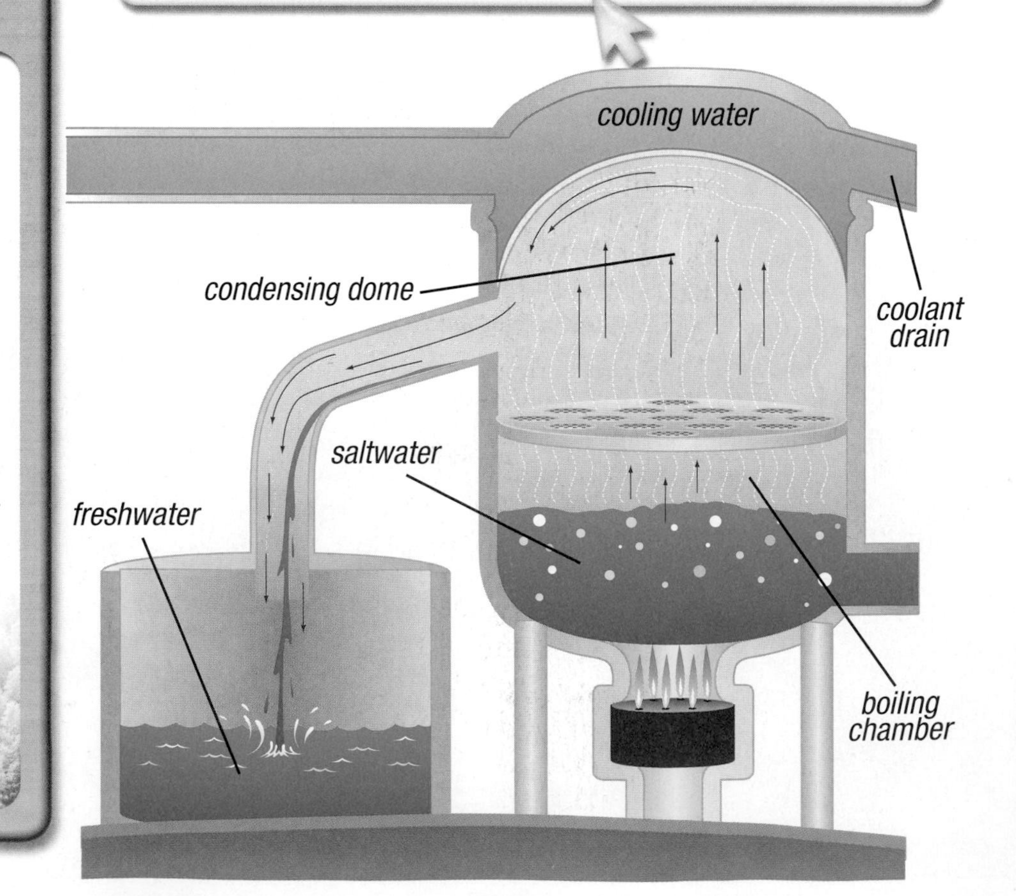

Some natural resources cannot be renewed. Nonrenewable resources either cannot be replaced or they are replaced much more slowly than they are consumed. Fossil fuels and minerals are common nonrenewable resources. Most scientists believe that fossil fuels like petroleum, coal, and natural gas are organic substances formed from the remains of once-living things. Minerals are naturally occurring inorganic substances found in the earth.

Humans, animals, and plants depend on soil, water, and minerals. Since natural resources are necessary for sustaining life, humans should act responsibly in conserving all of them. Water and air often become polluted. Large amounts of garbage are deposited into landfills. Decomposing garbage may seep into the nearby soil and water sources. Soil can also be contaminated by human waste, chemical fertilizers, and pesticides. Sometimes trees are harvested faster than they are renewed. Therefore, it is always important to make wise choices when using natural resources.

Household items such as batteries are thrown into landfills. Batteries contain poisonous metals, mercury, and lead, which seep into the ground and pollute it. Once these chemicals are in the soil, they are very difficult to extract.

The gasoline that powers automobiles is a nonrenewable resource.

Why did people in the 1700s start using coal instead of wood?

All living things rely on natural resources for survival. This baby gorilla eats plants for food. The plants must have soil, water, and minerals. Both animals and plants depend on the sun's energy and the air.

FOSSIL FUELS

VOCABULARY

coal ('kōl) a solid fossil fuel formed from decomposed plant remains

petroleum (pə·'trō·lē·əm) a crude oil, which is a liquid fossil fuel formed from microscopic plants, animals, and marine organisms

hydrocarbon ('hī·drō·kär·bən) a compound made of the elements hydrogen and carbon

natural gas ('na·chə·rəl 'gas) a mixture of methane and other gases formed from decomposed marine organisms

Roughly 80% of the world's energy consumption comes from fossil fuels. Fossil fuels form underground from decomposed plant and animal matter. Layers of sediment accumulate and trap the decayed remains. Over time, heat and pressure cause these layers to form fossil fuels.

Coal is a sedimentary rock that develops from the remains of plants that grew in swamps. When the plants died, they settled on the bottom of the swamp. As the plant remains built up, the partially decomposed matter formed peat—a dense, organic material. Eventually, the peat became covered with layers of sediment. Intense heat and pressure squeezed out all the water and air. Slowly the peat hardened, forming coal. Since coal is abundant and a relatively inexpensive fossil fuel, it is commonly used to produce electricity, steel, and cement. Coal is found throughout the world and is obtained by mining. The coal is then transported on trains and carrier ships.

In the past, what warned coal miners of gas leaks in the coal mines?

Coal is emptied from the mine cars and brought to coal tipples, or coal preparation plants. This is where the coal is sorted, weighed, sized, and placed onto trains for transport.

The world uses about 8,000 megatons (Mt) of coal per year. About 40% of that is used to fuel the world's electricity.

Petroleum is usually a thick, black liquid, often called *crude oil*. It formed from microscopic marine organisms such as plankton and algae. Land organisms also were part of its original material. Dead marine organisms settled on the ocean floor and decayed. Sediment built up, hardened, and turned into rock, burying the remains of the marine and land organisms. Over time, enormous heat and pressure caused the organic mixture to undergo physical and chemical changes. The mixture was broken down into compounds made of hydrogen and carbon, or hydrocarbons. Eventually, petroleum was formed. Petroleum is used for heating and making gasoline and jet fuel.

Natural gas also formed from decayed marine and land organisms. Petroleum and natural gas are often found together, but since gas is less dense than oil, it lies in rock layers above the oil. Natural gas and coal are also hydrocarbons. Most natural gas is used for heating and cooking. Gasoline, diesel, kerosene, and motor oil are made from petroleum. Both petroleum and natural gas are obtained by drilling.

What country drills the most oil?

QUICK FACT

Coal mining removes coal from the earth for use as fuel. If the coal is found near or on the surface, it is strip-mined. Most coal is deep underground and is extracted by using underground mining methods. Coal mining has been historically considered a rather dangerous job because of explosions, collapses, and respiratory diseases. Today, improvements in mining methods have reduced many of the risks of rock falls, explosions, and unhealthy air quality. Technological advancements have also made coal mining more productive than it has ever been.

Petroleum must be drilled from deep beneath the ground. Oil rigs like this one are commonly used to pump oil to the surface.

Oil is drilled from beneath the ocean by offshore drilling platforms.

MINERALS TO METALS

VOCABULARY

metal ('me·təl) a moldable substance that can reflect light and conduct heat and electricity

ore ('ôr) a naturally occurring rock from which useful metals or other minerals can be extracted

Minerals and metals are nonrenewable resources. Minerals occur naturally in the earth's crust and are made up of one or more chemical elements. About 99% of all minerals are made from the elements oxygen, silicon, aluminum, iron, magnesium, calcium, potassium, and/or sodium. There are more than 2,500 different minerals in all shapes, colors, and sizes. They can be identified by observing their physical properties such as hardness, color, luster, streak, and cleavage. Luster is the ability to reflect light. Streak is the color of the mineral when it is crushed and powdered. Cleavage is the tendency to break along smooth, flat surfaces.

Minerals form into unique geometric shapes called *crystals*. Some minerals or gemstones, such as diamonds, rubies, sapphires, and emeralds, are treasured because of their incredible beauty. Many gemstones are created when rocks in the earth's crust are buried, compacted, and heated by tectonic plate movement. Minerals are used in construction and in the making of glass and ceramics. Some minerals are found in foods and are necessary for the body's health.

Gemstones are minerals with exceptional beauty because of their color and crystalline structure. Amethysts and other minerals form inside geodes, which are hollow rocks lined with crystals.

Which mineral is sheetrock or drywall made from?

Some minerals are fluorescent and glow especially well under black light.

Strange but True

Would you eat this? Why not? This is Himalayan crystal salt, the cleanest and most beneficial salt in the world. It contains 80 minerals found in human bodies.

Some minerals are metals, a class of substances characterized by three main properties. Metals conduct heat and electricity. They can be bent or shaped into wire. Metals are generally shiny because they reflect light. Common metals include gold, silver, copper, lead, aluminum, iron, and platinum. Metals are a very important resource because people use them to build many kinds of things, including houses, bridges, automobiles, electronics, jewelry, and canned items.

Metals and some other minerals are removed from the earth through the process of mining. Ores are first extracted, or removed, from the ground. Valuable metals are removed from the ore and refined into useful materials. One metal that is very important is iron. Iron can be made into steel by removing most of its impurities. Steel is a stronger, more flexible metal used in buildings, cars, kitchen utensils, and many other manufactured products.

A tremendous amount of energy is needed to extract minerals from ore. Because minerals are nonrenewable, it is important to conserve them for the future.

How many known metals are there?

In open-pit mines, ore is mined downward in layers. The ore is then hauled by trucks for processing.

Steel is one of the main materials used in commercial buildings.

QUICK FACT

Minerals form into many different crystalline shapes, from cubes to elongated needles like this crocoite. The best specimens of crocoite are found on the island of Tasmania in Australia. Crocoite contains the elements lead, chromium, and oxygen. It is bright orange-red in color and has a streak of orange-red with a yellow tint. It has a translucent, shiny luster that is similar to a diamond.

REDUCE, REUSE, RECYCLE

VOCABULARY

landfill ('land·fil) a site for the disposal of waste materials

compost ('käm·pōst) a mixture that consists largely of decayed organic matter and is used to fertilize soil

Have you ever thought about how much trash you throw away and where it goes? On average an American throws away over 1.8 kg (4 lb) of trash a day, or 657 kg (1,460 lb) each year. With over 300 million Americans, the amount of waste is a big concern. Garbage is hauled off by trucks and placed in landfills. Modern landfills are lined on the sides and the bottoms with layers of clay or plastic to help keep the toxins from leaking into the soil. When the landfill is full, it is sealed and covered with a cap of clay and dirt.

One of the best ways to lessen the need for more landfill space is to reduce, reuse, and recycle. Reduce the amount of waste by purchasing items that do not have multiple layers of packaging material. Reuse items instead of throwing them away. Recycle by taking plastic, cans, glass, and paper to recycling centers.

Much of your trash can be recycled. The items to be recycled can be taken to recycling centers or picked up by waste management companies.

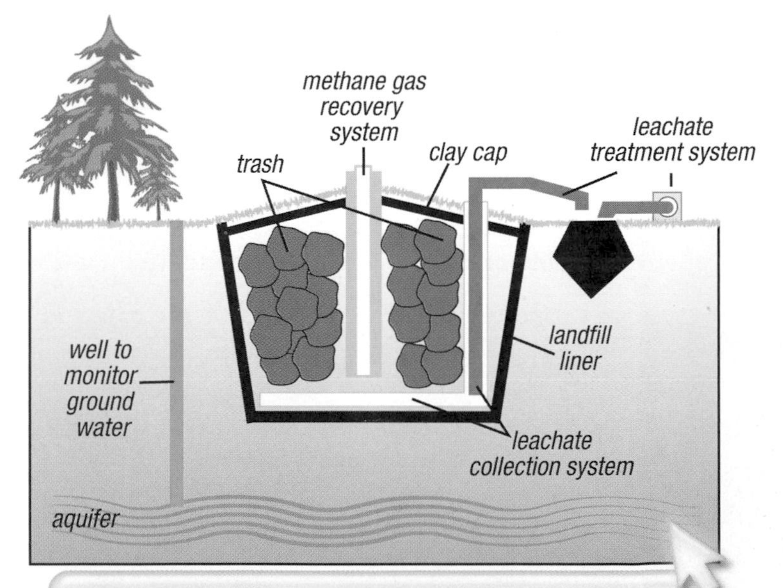

Today's landfills are very different from the open dumps in the past. Modern landfills are situated near clay deposits to protect the surrounding environment. Landfills are covered on the bottom and the sides with special liners. Wells are drilled in order to detect contamination in the water supply.

Recycle 31%
Landfill 55%
Burn 14%

About 55% of the trash in the United States is taken to landfills. The other 45% is either recycled or burned. A new landfill costs millions of dollars to build.

Making compost is one way to reduce, reuse, and recycle at the same time. Compost is a mixture of organic substances that is used to fertilize soil. Compost piles are made by mixing leaves, flowers, grass clippings, weeds, coffee grounds, and fruit and vegetable peels together with water and nitrogen. The mixture can be put into boxes or bins where it will be ready in four months or less, depending on what is in it and whether it is mixed up every 4–5 weeks. Mature compost is called *humus*. People put this around plants, flowers, and gardens to add more nutrients to the soil.

Think about some of the things that get thrown away. What is being placed into the trash that is hazardous to the environment? Paint, paint thinners, household cleaners, brake and transmission fluids, aerosol cans, pesticides, and batteries are just a few of the items that are toxic. Instead of throwing them away, they should be properly disposed of by taking them to household hazardous waste stations.

When your family gets a new computer, what happens to the old one?

QUICK FACT

Millions of tires are discarded each year. Huge stockpiles of tires create areas for mosquitoes to live and breed. Sometimes, because of lightning or other reasons, the tires catch on fire, which produces tremendous air pollution. Rubber tires can be successfully recycled. The rubber is ground into small, crumb-like pieces, which are then used to make other rubber products. Tires with a good amount of tread remaining on them can be recycled by being retreaded. Retreading tires reduces the need for manufacturing new tires.

To speed up the production of compost, gardeners occasionally stir the mixture, after adding nitrogen-laden fertilizer and water.

Earthworms compost garbage faster than any other method. They can eat at least one-third their weight in food daily.

ALTERNATIVE FUEL SOURCES

VOCABULARY

biomass (ˈbī·ō·mas) the organic matter from plant material and animal waste used as a source of energy

geothermal energy (jē·ō·ˈthər·məl ˈe·nər·jē) the energy produced by heat trapped within the earth

Today's society is dependent on fossil fuels because they are easily accessible and relatively inexpensive. Fossil fuels are nonrenewable natural resources that may not always be available. To continue to have access to energy, it is necessary to find and use alternative energy resources that are renewable. Solar energy, biomass, and geothermal energy are examples of these alternative sources.

Most forms of energy originally come from the sun. The sun supplies more than enough solar energy to meet the world's energy needs and it is a renewable resource. Solar cells located in solar panels absorb sunlight and convert it into electrical energy. People use this energy to heat homes and supply electricity to buildings. Solar collectors are flat, rectangular boxes used to heat water. The boxes are often painted black and have black piping that water runs through. The water receives energy in the form of heat and is then pumped into storage tanks.

Solar collectors absorb heat energy from the sun. As water flows through the pipes inside the collectors, the heat energy warms the water.

Solar cells located on solar panels capture the sun's energy, which is converted into electrical energy.

Solar powered calculators reduce the use of batteries because they obtain their energy from the sun. Calculators and watches need only a few solar cells, while it takes hundreds of solar cells within solar panels to provide electricity for a house.

Plants absorb the sun's energy and convert it into food energy that is stored in the leaves, stems, and roots. Animal dung from plant-eating animals also has stored energy. These sources of energy are referred to as biomass. The stored energy in biomass can be used to produce fuel. Energy is usually released from the biomass by burning it. Wood and charcoal are common examples of biomass and are burned for heating and cooking. In areas where firewood is scarce, animal dung is dried and burned as an energy source.

Geothermal energy is produced from heat trapped within the earth. In some areas groundwater comes close to a source of magma and is heated by it. This produces steam. Geothermal power plants harness the energy from this steam and use it to generate electrical energy or to heat water. The hot water can then be piped through buildings for heat. Geothermal power plants in California can supply electrical energy for 2–4 million homes!

LINKS

Writing Link
Write a letter to your local state representative and encourage him or her to direct funds toward research for alternative fuel sources. Include what type of alternative sources you think should be researched and why you think they would benefit the environment.

Research Link
Research Christian organizations that focus on helping people care for their environment or helping them get access to clean water. Also research how solar powered items can be used for evangelism in some remote areas of the world. Present your findings to the class.

Communication Link
Interview a person who sells or owns a hybrid or electric car. Ask how the car works; how much gas it uses, if any; how often the batteries need to be charged; and how it is better for the environment. Write an essay about what you have learned.

Magma can heat groundwater to temperatures over 148°C (300°F). Hot water and steam escape through geysers or natural vents in the earth's crust.

CHAPTER 11

WEATHER AND CLIMATE

An envelope of gases surrounds the earth. These gases form the atmosphere, which is divided into layers by temperature. The atmospheric layer closest to Earth is called *the troposphere*. This layer, which extends upward approximately 12 km (7.5 mi), is where most of the earth's weather occurs. Weather is affected by the sun's energy.

About 50% of the sun's energy is absorbed by the earth's surface. The energy that is not absorbed is reflected back into the atmosphere. As air is heated it becomes less dense, rises, and creates areas of low air pressure. Cooler, denser air moves under the warmer air, creating areas of high pressure. This movement of air is referred to as *convection currents*. These currents cause wind.

VOCABULARY

weather ('we·thər) the condition of the atmosphere at any given moment in a particular area

wind ('wind) the movement of air caused by differences in air pressure

KEY IDEAS

- Air temperature, air pressure, and humidity determine weather.
- Convection currents cause local and global winds.
- The interaction of air masses forms fronts and causes changes in the weather.

QUICK FACT

The greenhouse effect is a naturally occurring process that insulates and warms the earth. It keeps the earth's atmosphere at a temperature that is suitable for life. Some of the sun's radiation is absorbed by the earth's surface and is converted into infrared radiation, or thermal energy. This thermal energy is released back into the atmosphere. Greenhouse gases in the atmosphere, such as water vapor, carbon dioxide, and methane, absorb the thermal energy. This causes more energy to be added to the atmosphere. The heat energy in the atmosphere is once again absorbed by the earth's surface. Then it is radiated back into the atmosphere. This cycle helps keep the earth and its atmosphere at a temperature comfortable for most living things.

Strange but True

Is this a rainbow?

VOCABULARY

sea breeze ('sē 'brēz) a local wind that blows from an ocean or sea toward the land

land breeze ('land 'brēz) a local wind that blows from the land toward a body of water

QUICK FACT

The state of Florida in the United States is a peninsula surrounded by the Atlantic Ocean and the Gulf of Mexico. No matter which direction the wind blows, it is always from the water. Sea breezes from both coasts can collide in the middle, creating severe storms. Sometimes these storms even bring hail because of the tremendous uplift that is occurring in the atmosphere. Because of this, Florida gets struck by lightning more than any other state in the United States and more than most other places in the world.

Most differences in air pressure are caused by the unequal heating of the earth. Energy from the sun penetrates through the layers of the atmosphere and heats the earth's surface. Some surfaces absorb more energy than others. In addition, the equator receives more direct sunlight than other latitudes. Air over the earth expands as it is heated, becomes less dense, and rises. The rising of warmed air is called *uplift*. This uplift and the convection currents that result from it create winds. Winds that occur over short distances are called *local winds*.

Unequal heating often occurs near large bodies of water. During the day the land heats up faster than the water. This causes the air over the land to become warmer than the air over the water. The warmer air over land rises, while the cooler air that is over the ocean comes in to take its place. This pulling of cool air from the water that replaces the rising warmer air over the land creates a local wind called a sea breeze.

Does a sea breeze only occur by the sea?

During the day, the cooler air over the water blows in and replaces the rising warmer air over the land. This movement of air from water to land is called *a sea breeze*.

Sea Breeze

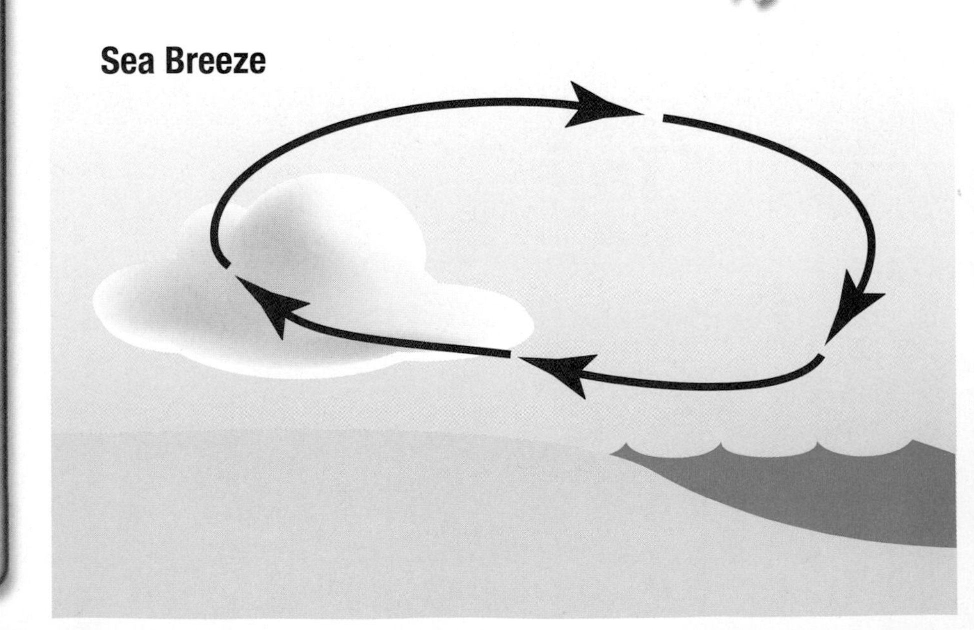

At night the land cools down faster than the water. If the land cools below that of the water, the air over the water becomes warmer than the air over the land. This causes the air to rise and creates lower pressure over the water. The cooler air from the shore has higher pressure and begins to flow down toward the water, replacing the warm air. This movement of air from land to water creates a local wind called a land breeze.

If a land breeze develops, it will occur from late night to early morning. Once the land heats up again in the morning, the land breeze decreases. If the air over the land gets warmer than the air over the water, a sea breeze develops. Sea breezes are most common along tropical coasts. They occur most often in midsummer during the day and early evening when there are large temperature differences. Land breezes occur less frequently. They tend to develop along coastlines with steep shorelines or in the temperate regions during the colder months.

At night the land cools off more quickly than the water. The air over the water is now warmer than the land. The cooler air over the land blows toward the water to replace the rising warmer air, causing a land breeze.

Land Breeze

IN THE FIELD

Wind direction is described by the direction from which it blows. A northerly wind blows from the north to the south. A westerly wind blows from the west to the east. Scientists use various instruments to measure wind speed and direction. Anemometers are used to measure wind speed. Weather vanes and windsocks are used to measure wind direction. Special instruments called *rawinsondes* measure both wind speed and wind direction. They are attached to weather balloons and are designed to measure air pressure, altitude, temperature, humidity, and geographical position, as well as wind speed and direction.

Sand can be used to help see which way the breeze is blowing.

GLOBAL WINDS

VOCABULARY

doldrums ('dōl·drəmz) a relatively calm area near the equator where warm air rises

horse latitudes ('hôrs 'la·tə·to͞odz) the relatively calm areas located near both 30° latitudes where cool air sinks

trade winds ('trād 'windz) the winds that blow easterly from both 30° latitudes toward the equator

westerlies ('wes·tər·lēz) the winds, found between the 30° and 60° latitudes, that blow toward the poles from west to east

polar easterlies ('pō·lər 'ē·stər·lēz) the cold winds from the poles to the 60° latitudes that blow from east to west

Unequal heating of the earth creates temperature differences between the equator and the poles. These variations in temperature cause changes in air pressure and produce giant convection currents in the atmosphere. The warmer air at the equator rises and flows toward the poles. The colder air near the poles sinks and flows toward the equator. As a result the air pressure at the poles is higher than it is around the equator. This difference in air pressure causes the movement of air over large areas, which is called *global winds*.

The earth rotates on its axis from west to east, or counterclockwise. This affects the direction of global winds. The curve of the winds and the ocean currents because of the earth's rotation is known as *the Coriolis effect*. In the Northern Hemisphere, northbound wind currents will bend toward the east and southbound currents will bend toward the west. In the Southern Hemisphere, the opposite occurs.

The yellow arrows indicate the wind if there were no Coriolis effect. The burgundy arrows indicate the approximate wind patterns caused by the Coriolis effect.

The Coriolis effect has little to no effect on how water goes down the drain.

The Coriolis effect and differences in air pressure combine to form a pattern of wind belts and calm areas around the earth. Near the equator is a calm area of low pressure called the doldrums. At latitudes of about 30ºN and 30ºS is the calm areas of high pressure called the horse latitudes. Both of these areas experience little to no wind. Between the doldrums and the horse latitudes are wind belts called the trade winds. The cool air in these regions sinks and blows toward the equator.

The wind belts found between the 30º and 60º latitudes flow toward the poles and are called the westerlies. They often bring moist air and precipitation. Between the 60º and 90º latitudes in both hemispheres are the polar easterlies. The air here is cold and sinks toward the lower latitudes. When the cold air of the polar easterlies meets the warmer air of the westerlies, the weather of the countries in those areas is greatly affected.

Can you tell where this yacht might be located?

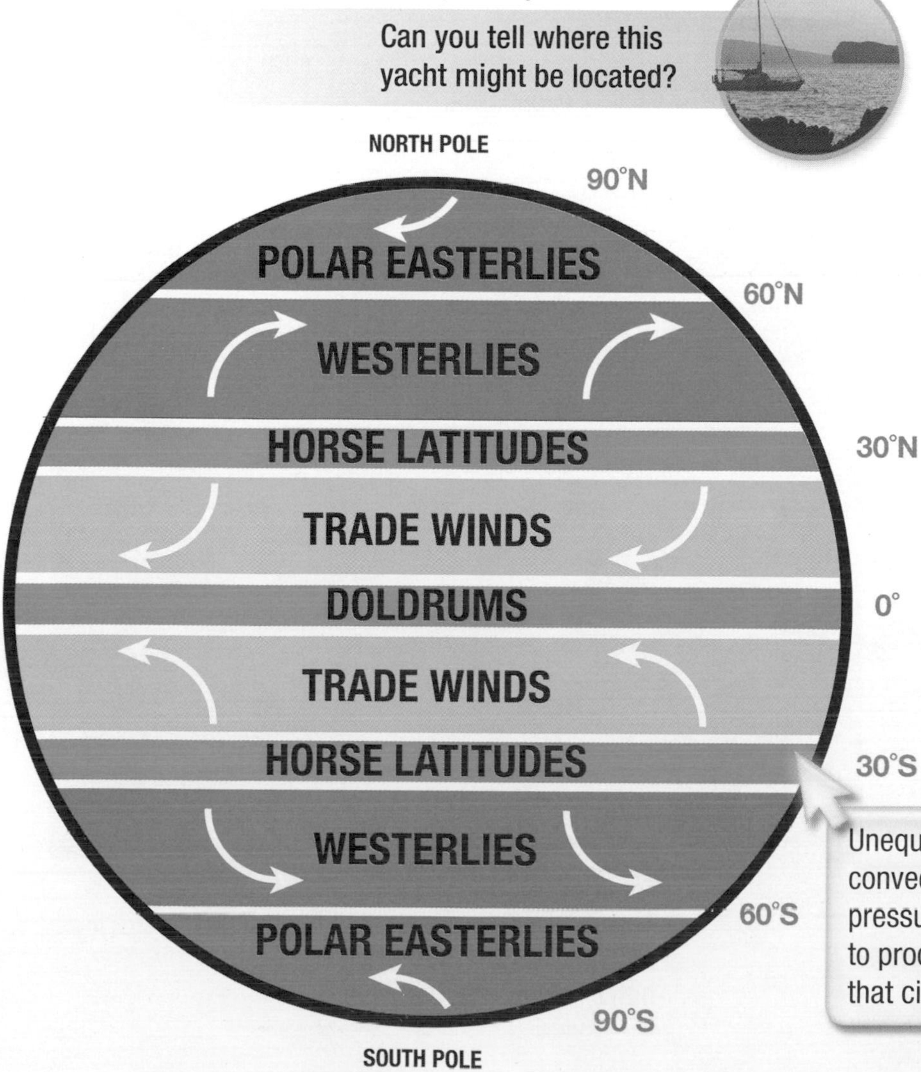

Unequal heating of the earth's surface, convection currents, differences in air pressure, and the Coriolis effect combine to produce wind belts and calm areas that circle the earth.

QUICK FACT

Legend has it that centuries ago ships sailing to the New World often got stranded in the calm regions of the horse latitudes. Supplies would run short and the sailors were faced with a huge decision. They had to figure out how to make the food and water last. Often their decision was to throw their cargo of horses, now dead or dying of thirst or starvation, overboard to conserve water and food. Could this be how the horse latitudes got their name?

MASSES AND FRONTS

VOCABULARY

air mass ('âr 'mas) a very large body of air that maintains uniform temperature, humidity, and pressure as it moves

front ('frunt) the boundary at which two different air masses meet

Global winds move large bodies of air called *air masses*. An air mass is a very large collection of air that has uniform temperature, humidity, and air pressure. It develops slowly and takes on the characteristics of the region in which it originates. A body of tropical air forms in the tropics where it is warm. A body of polar air forms north of latitude 50ºN and south of latitude 50ºS and is cold. An air mass which develops over water is humid and referred to as *maritime*. Bodies of air that form over land are called *continental* and tend to be dry.

Air masses are classified according to both temperature and humidity. Therefore, an air mass that forms over warm oceans is called *maritime tropical*. When a body of air forms over cold oceans, it is called *maritime polar*. *Continental tropical* masses form over warm land and ones that are *continental polar* develop over cold land.

Where would a continental tropical air mass form?

Maritime tropical air masses form over large bodies of warm water such as the Gulf of Mexico and the tropical Atlantic. They have warm, humid air and often bring precipitation.

Continental polar air masses form over regions such as the Canadian Rockies. They have cold, dry air. The movement of continental polar air masses can bring in bitter cold during the winter.

As air masses move across the earth, they tend to collide with each other. They do not blend or mix easily because they have different temperatures and densities. When air masses meet, they form a boundary called a front. Fronts often bring a change of weather, such as precipitation and storms.

One of the most common fronts is a warm front. It is the zone where faster-moving warm air collides with a body of cold air. The cold air is more dense, so the warm air rises up and over the cold air. If the warm air is humid, precipitation occurs along the front. If it is dry, scattered clouds usually form.

Another common front is a cold front. This occurs when an advancing cold air mass collides with slow-moving warm air. The cold, dense air sinks and moves under the warm air. The rising warm air is pushed upward where it cools and expands. If the warm air is humid, the water vapor condenses, clouds develop, and precipitation is likely to occur.

What kind of front develops when cold and warm air masses meet, but neither overtakes the other?

The boundary between a rapidly moving cold air mass and a slowly moving warm air mass is called *a cold front.* After a cold front passes through an area, colder, dryer air moves in and brings clear skies.

cold

warm

A warm front develops when an advancing warm air mass overtakes a slowly moving cold air mass. After a warm front moves through an area, the weather is likely to be warm and humid.

warm

cold

OUTLOOK

Weather affects people, but people can affect weather too. When cities are constructed, asphalt and concrete surfaces transmit the heat they absorb, raising temperatures even higher. Drain and sewer systems quickly carry off water, making the area drier. Urban and industrial areas contribute air pollutants that linger in areas where the winds are light. This can create health hazards. Air pollutants can also cause acid precipitation. What is commonly viewed as progress can sometimes be seen as the opposite. Although modern cities can affect weather patterns, severe weather dates back to the Flood (Genesis 7) and will continue through the Second Coming, Jesus Christ's return to Earth (Revelation 6:1–16:21).

Strange but True

This is actually not a rainbow! It is a rainbow-like phenomenon called *a circumzenithal arc.* For this to occur, ice crystals in the clouds must be arranged horizontally so that they act like 90° prisms. This arrangement makes the sunlight bend in such a way that it forms this unusual arc.

WEATHER MAPS

VOCABULARY

isobar ('ī·sə·bär) a line used on a map or chart connecting points of equal air pressure

QUICK FACT

Norwegian physicist and meteorologist Vilhelm Bjerknes is considered to be one of the founders of modern meteorology and weather forecasting. In 1917 he founded a school of meteorology for young meteorologists, including his son, Jacob. Vilhelm suggested the term *front* to signify the boundary where two air masses meet. He chose this term because the collision that occurs reminded him of a World War I battlefront during a military operation. In honor of his achievements, the Bjerknes crater on the moon and a crater on Mars were named after him.

Since weather affects people by influencing their working and living conditions, clothing choices, and traveling schedules, weather prediction is very important. Scientists use many different types of instruments to collect data that helps them forecast the weather. A weather forecast is a prediction of weather conditions over the next three to five days. A meteorologist is a person who observes, collects, and analyzes data on the atmospheric conditions and then uses the data to make weather predictions. Data from local weather stations is usually reported to national weather services, which then assemble the information for national and worldwide forecasts.

Weather maps can be found on the Internet, in newspapers, and on television weather reports. Special symbols are placed on weather maps to represent specific data. It is important to be able to recognize these symbols in order to interpret the weather information.

What is this symbol?

One type of instrument used by scientists is this meteorology sphere. It collects information about atmospheric conditions.

Weather stations are located around the world. This one is on a mountaintop in the Alps of Italy.

Standard symbols on weather maps indicate different types of weather conditions. Isobars are lines that show areas with similar air pressure. When they form closed circles, they indicate high or low pressure areas. A capital *H* can represent a center of high pressure, while a capital *L* stands for a center with low pressure. Various symbols are used to represent types of precipitation. Station models convey temperature, cloud cover, and wind speed and direction.

A cold front is often represented by a solid blue line with triangles attached to the line pointing in the direction that the cold front is moving. A warm front can be represented by a solid red line with semicircles attached to the line pointing in the direction that the warm front is moving.

Improvements in technology have made the accuracy of two-to-three-day forecasts possible. Meteorologists also make extended forecasts, which are predictions beyond five days. These, however, are less accurate because small changes in weather can lead to large changes overall.

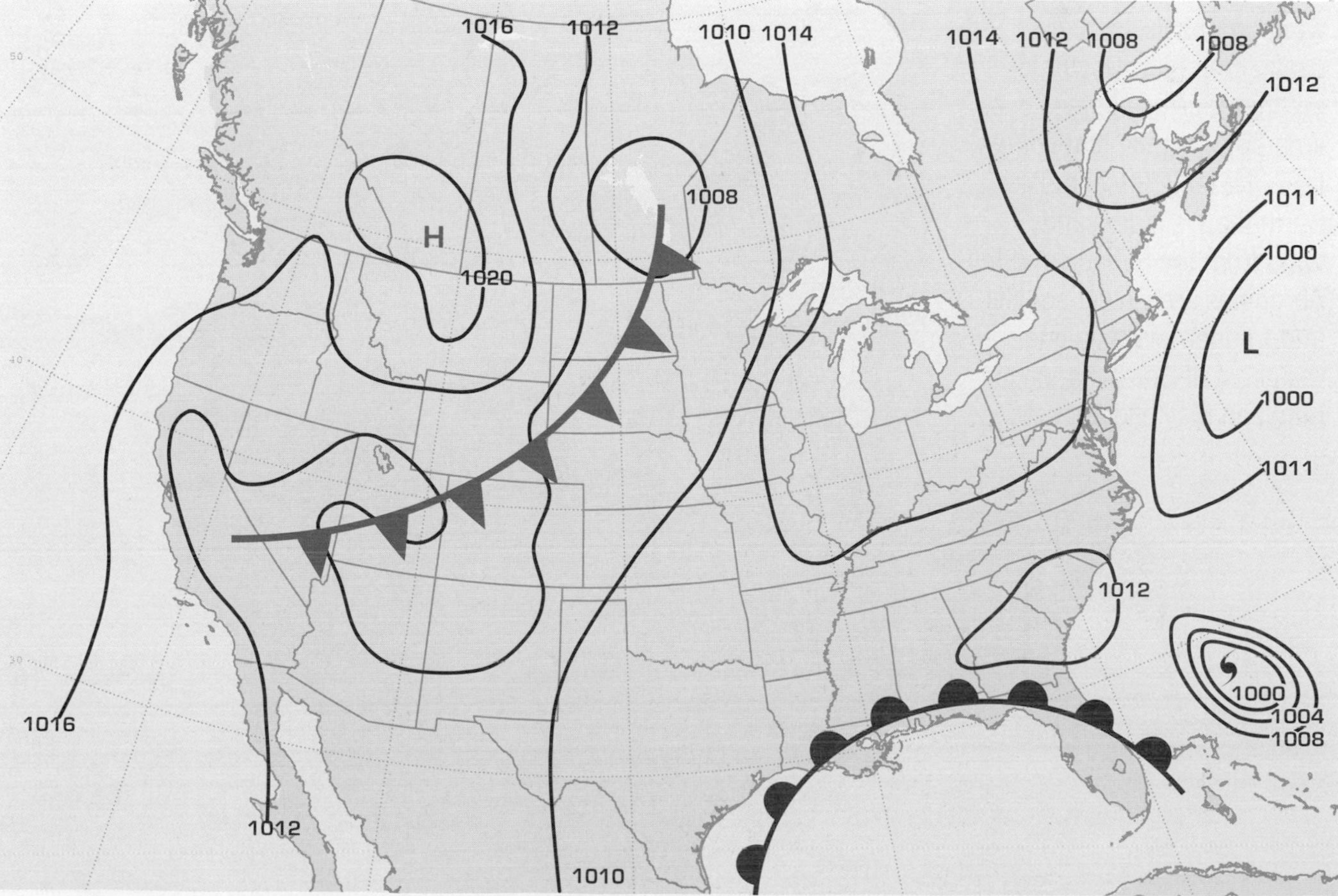

VOCABULARY

monsoon (män·ˈsōōn)
a wind that changes direction with the change of seasons and which usually brings very heavy rainfall

QUICK FACT

Many local winds have their own special names. Alizé Maritime is a wet, fresh northerly wind that blows across west central Africa. The Bayamo is a violent wind on Cuba's southern coast. Fremantly Doctor is an afternoon sea breeze over Western Australia that comes from the Indian Ocean. The Harmattan is a dry sandy wind from the Sahara, and the Abroholos are violent squalls and frequent winds along the coast of Brazil that occur between May and August.

An area's landform can also influence local winds. If the landscape is mountainous, it can be prone to certain local winds called *valley* and *mountain breezes*. When the sun rises, the mountain peaks are the first to receive light. As the day progresses, the rocks and dirt on mountain slopes absorb energy more quickly than the plants in the valleys. This causes the mountain slopes to heat up faster than the valleys nearby. As the warm air rises off the slopes, the cooler air from the valley rushes in to replace it. This upslope wind is called *a valley breeze*.

The valley continues to heat up during the day and by late afternoon it radiates more heat than the mountain slopes. Mountains can lose heat quickly if there is a lack of dense vegetation to hold in the heat. The cool air from the mountain slopes sinks down toward the valleys and pushes the rising warm air upward creating *a mountain breeze*. This is why the temperature drops rapidly on a mountain after sunset.

What is the downwind side of the mountain called?

Another type of wind that blows around the world is the jet stream. When cold, polar air meets the warmer air of the middle latitudes, it creates extreme differences in pressure. This forms strong, high-speed winds that blow in the upper troposphere and lower stratosphere. These winds generally flow above global winds and from a westerly direction. They can reach speeds up to 400 kph (248 mph)! Airplane pilots often take advantage of jet streams to save time and fuel.

Seasonal winds occur in various areas throughout the world. The air over the land during the summer months is warmer than the air over nearby bodies of water. In the winter the opposite is true. The air over the water is warmer than the air over the land. The monsoons are an example of seasonal winds that change direction with the change of seasons. The monsoons are most known in eastern and southern Asia, but can also be observed in Australia, West Africa, the Pacific Ocean, South America, and even in southwestern North America.

What kind of weather do the summer monsoons usually bring?

Monsoons occur between July and September in southwestern North America. This storm is a result of the effect of the monsoons in Arizona.

In the Philippines monsoons often bring downpours of rain during the midday.

LINKS

English Link
Choose and read a book about native folklore that includes weather. Write a book report and present it to the class. Draw or show pictures from the book during your oral report. Discuss the relevance of weather conditions to the story.

History Link
Create a timeline on the history of weather instruments. Research the inventions of the barometer, hygrometer, thermometer, rain gauge, telegraph, weather map, and satellite. Include them on your timeline. Present your timeline to the class.

Bible Link
Use a concordance or topical index to find Scriptures that include references to the weather. Look up the verses. Prepare and give a devotional to the class based on your findings.

CHAPTER 12

SUN, EARTH, AND MOON

KEY IDEAS

- Earth's seasons occur because of the tilt of the earth on its axis and its revolution around the sun.
- The earth and moon experience two types of movement—rotation and revolution.
- Gravity and inertia keep Earth orbiting around the sun and the moon orbiting around Earth.
- The relative positions of the sun, the earth, and the moon produce eclipses, tides, and moon phases.

The positions of the sun, the earth, and the moon determine the earth's seasons and moon phases. These positions also influence lunar and solar eclipses, as well as high and low tides.

Earth and its moon experience two different types of motion as they move through space. The earth spins on its axis once every 24 hours, or one solar day. As a result, this planet experiences night and day. The moon also spins on an axis about once every 29.5 days, or one lunar day. Both axes are tilted, although the moon's is tilted significantly less than the earth's.

Another similarity between the earth and the moon is that they both revolve around another object. Earth revolves around the sun once every 365.25 days. The moon revolves around Earth counterclockwise approximately every 29.5 days. This is the same amount of time it takes for the moon to rotate once on its axis. Both Earth and its moon follow elliptical paths, or orbits, as they complete their revolutions.

QUICK FACT

Where you are on Earth determines how fast you are moving. This is due to the earth's rotation. For example, if you were standing on the equator, you would be moving at a speed of about 1,609 km (1,000 miles) per hour because of Earth's rotation.

Strange but True

What solar system object is this?

VOCABULARY

law of universal gravitation ('lȯ 'əv yo͞o·nə·'vûr·səl gra·və·'tā·shən) every object in the universe attracts every other object, depending on their masses and distances from one another

The English scientist Sir Isaac Newton observed an apple falling from a tree to the ground. Based on his observations, he hypothesized that the same force acting on the apple was causing the moon to stay in its orbit around Earth. Once Newton realized that gravity occurs everywhere, he proposed the law of universal gravitation. This law states that every object in the universe attracts every other object. The force of attraction between two objects is determined by their masses and the distance between them.

The greater the mass of an object, the greater the gravitational pull of that object. Gravitational pull also increases as the distance between two objects decreases. Therefore, as the distance between objects increases, the gravitational force between them decreases. Since weight is determined by the force of gravity, an object's weight varies depending on where it is in the universe. The mass of an object is the same wherever it is located.

IN THE FIELD

The GRACE (Gravity Recovery and Climate Experiment) mission is a joint project of the United States and Germany. The goal of this mission is to map variations in Earth's gravity field using two spacecraft. The satellites were launched in 2002. In the first 10 years of orbit, they circled Earth more than 55,000 times.

The force of gravity from the moon and the earth work together to keep the moon in orbit.

When masses increase, the force of gravity increases. When the distance decreases, the force of gravity increases.

m m m m 2m 2m

The sun's mass is 330,000 times that of Earth, and the sun is closer to Earth than any other star. Therefore, it exerts an enormous gravitational pull on the earth, keeping the planet revolving. Since Earth is massive compared to its moon, the earth's gravitational pull on the moon is greater than the moon's gravitational pull on Earth. However, compared to natural satellites that orbit other planets, the moon is huge. Its relatively large mass, combined with its nearness to Earth, means that the moon exerts a significant gravitational force on the earth. This force influences some of the planet's processes.

Inertia is another factor that keeps the earth orbiting around the sun and the moon orbiting around the earth. Newton's first law of motion states that objects at rest will stay at rest and objects in motion will stay in motion unless acted upon by an outside force. Earth stays in motion because of its inertia.

Are rotation and revolution the only types of motion that the earth experiences?

QUICK FACT

The earth's rotation is slowing down approximately two milliseconds per century. As a result, the moon's orbit moves about 3.8 cm ($1\frac{1}{2}$ in.) farther away from Earth each year. Both of these changes are effects of the complex gravitational forces that act upon Earth and its moon.

Earth's mass is not spread out evenly. As a result, Earth's field of gravity is not uniform either. This world gravity model is exaggerated. However, its lumpiness shows the unequal distribution of Earth's mass.

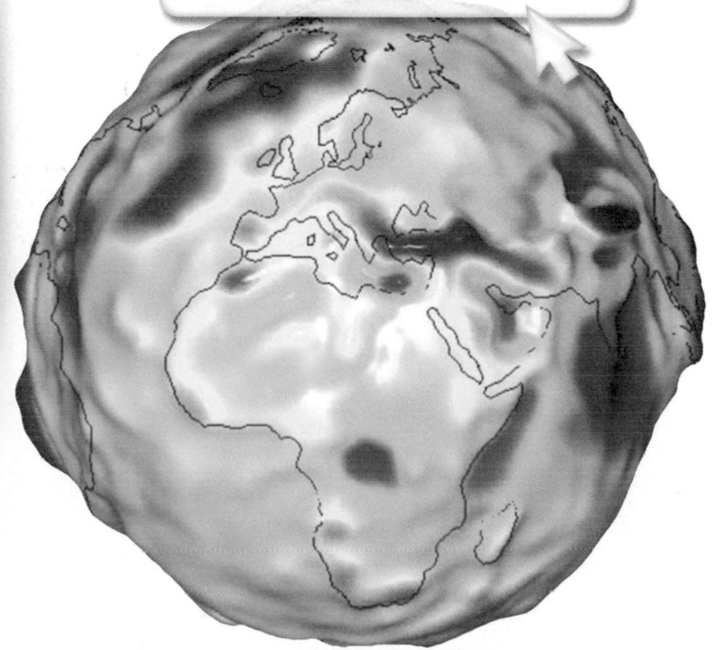

The sun contains almost all of the mass in the whole solar system! Its strong gravitational pull keeps Earth orbiting around it year after year. Notice the large solar flare erupting from the sun's surface.

SEASONS

VOCABULARY

solstice ('säl·stəs) the two days of the year in which the sun's most direct rays reach farthest north or farthest south

equinox ('ē·kwə·näks) the two days of the year when day and night are almost the same length everywhere on Earth

As Earth revolves around the sun, many areas experience four seasons. Spring, summer, fall, and winter are characterized by variable temperatures, weather patterns, and length of daylight. This change of seasons is caused by the tilt of the earth as it rotates on its axis and revolves around the sun. The planet's tilt affects where the sunlight shines on Earth. For example, sunlight shines on Earth most directly at the equator and least directly at the North and South Poles.

The people living near the equator do not experience four distinct seasons. Instead, they have alternating wet and dry seasons. Those living near the North and South Poles have short summers and long winters. There the sun provides light for 24 hours each day for most of the summer season, although the sun remains low in the sky. During the polar regions' winter season, it is continually dark for months!

What is the midnight sun?

SPRING
Neither of Earth's hemispheres is tilted toward the sun during the spring equinox.

WINTER
The Southern Hemisphere is tilted toward the sun. The winter solstice occurs.

SUMMER
The Northern Hemisphere tilts toward the sun. The summer solstice occurs.

FALL
Neither of Earth's hemispheres is tilted toward the sun during the fall equinox.

On or about June 22, people in the Northern Hemisphere experience more daylight hours of sunshine than any other day of the year. This is known as the summer solstice. Around December 22 the opposite occurs during the winter solstice, when there are more hours of darkness than daylight. These solstices take place because of the maximum tilt of the Earth's axis toward and then away from the sun.

Equinox occurs when neither of Earth's poles is close to the sun. In the Northern Hemisphere the spring equinox happens around March 21. The fall equinox occurs around September 22. During these events the sun appears to be directly above the equator and the amounts of daylight and nighttime are almost equal everywhere on Earth. The Southern Hemisphere's spring and fall are opposite of the Northern Hemisphere's.

OUTLOOK

Many people incorrectly think that the seasons are caused by how far Earth is from the sun. However, since Earth's orbit is only slightly elliptical, almost circular, its distance from the sun varies by about 3% throughout the year. This amount of difference is not enough to fully account for the seasonal solstices and equinoxes. The angle of the sun's rays, affected by the tilt of Earth's axis, is a more important factor. The Bible says that God made day and night and summer and winter (Psalm 74:16-17).

What is unique about the equator?

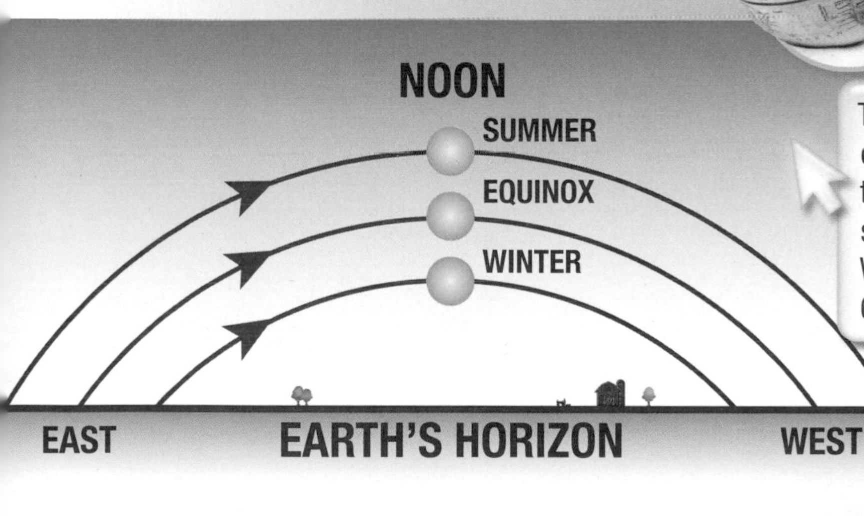

The height of the sun from Earth's horizon changes throughout the year. It is highest in the Northern Hemisphere during the summer solstice and lowest during the winter solstice. When both equinoxes occur, the sun is about 60° above the horizon at noon.

Strange but True

This is the moon. NASA's *Galileo* space probe took 53 different images of it. Colors were added to this picture to highlight certain lunar features. The pinkish areas are mountains. Volcanic lava flows are shown in blue to orange shades. The dark blue areas show areas that are richer in the metallic element titanium than the green areas.

PHASES AND ECLIPSES

VOCABULARY

crescent ('kre·sənt) the moon phase in which less than half of the moon's sunlit side is visible

gibbous ('ji·bəs) the moon phase in which more than half, but not all, of the moon's sunlit side is visible

wax ('waks) to grow

wane ('wān) to shrink

lunar eclipse ('lo͞o·nər i·'klips) an event that occurs when Earth passes directly between the sun and the moon, causing Earth's shadow to block the sun's light from the moon

solar eclipse ('sō·lər i·'klips) an event that occurs when the moon passes directly between the sun and Earth, causing the moon's shadow to block the sun's light from a portion of Earth

As the moon revolves around Earth and rotates on its axis, people see portions of its sunlit side. The shapes indicate which phase the moon is in. Each phase results from the positions of the earth, the moon, and the sun. Eight main phases are the new, first quarter, full, and third quarter moons, as well as two crescent and two gibbous moons. A crescent moon is similar in shape to the letter C. A gibbous moon is larger than a semi-circle but is not a complete circle of light. When the moon is directly between Earth and the sun, it is called *a new moon*, which cannot be seen. As the moon travels counterclockwise around Earth, it appears to grow, or wax. When all of the near side is illuminated, it is called *a full moon*. After a full moon occurs, it appears to shrink, or wane, as it returns to a new moon. The moon's near side always faces Earth while most of its far side faces away from Earth.

When can the moon be seen on Earth?

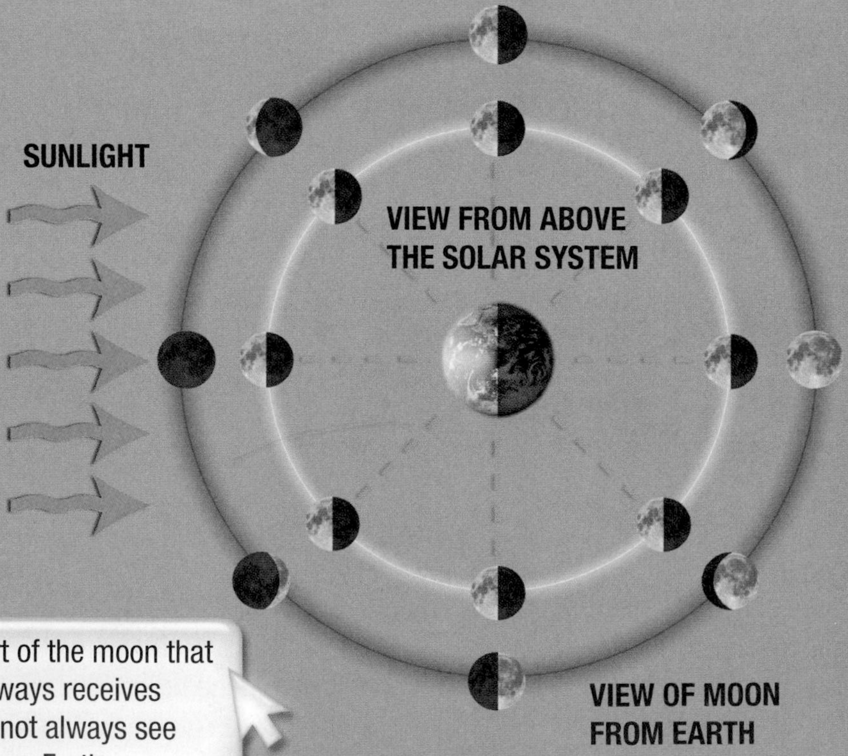

Although the part of the moon that faces the sun always receives sunlight, we cannot always see the sunlit side from Earth.

When the earth and moon are in line with the sun, they occasionally block sunlight from each other. These events are called *eclipses*. There are two types of eclipses—lunar and solar. A lunar eclipse occurs when Earth blocks the sun's light from the full moon. For this to happen, Earth must be between the sun and the moon. A solar eclipse blocks sunlight from a portion of the earth. This takes place when the moon passes directly between the sun and the earth during a new moon phase. The moon's color and brightness during an eclipse varies because of the amount of light refracted or bent by the earth's atmosphere.

QUICK FACT

During a total lunar eclipse, the earth blocks sunlight from the moon. However, some light from Earth's atmosphere is reflected by the moon, causing the moon's color to range from gray to a copper color.

During a total solar eclipse, the moon blocks sunlight from the Earth.

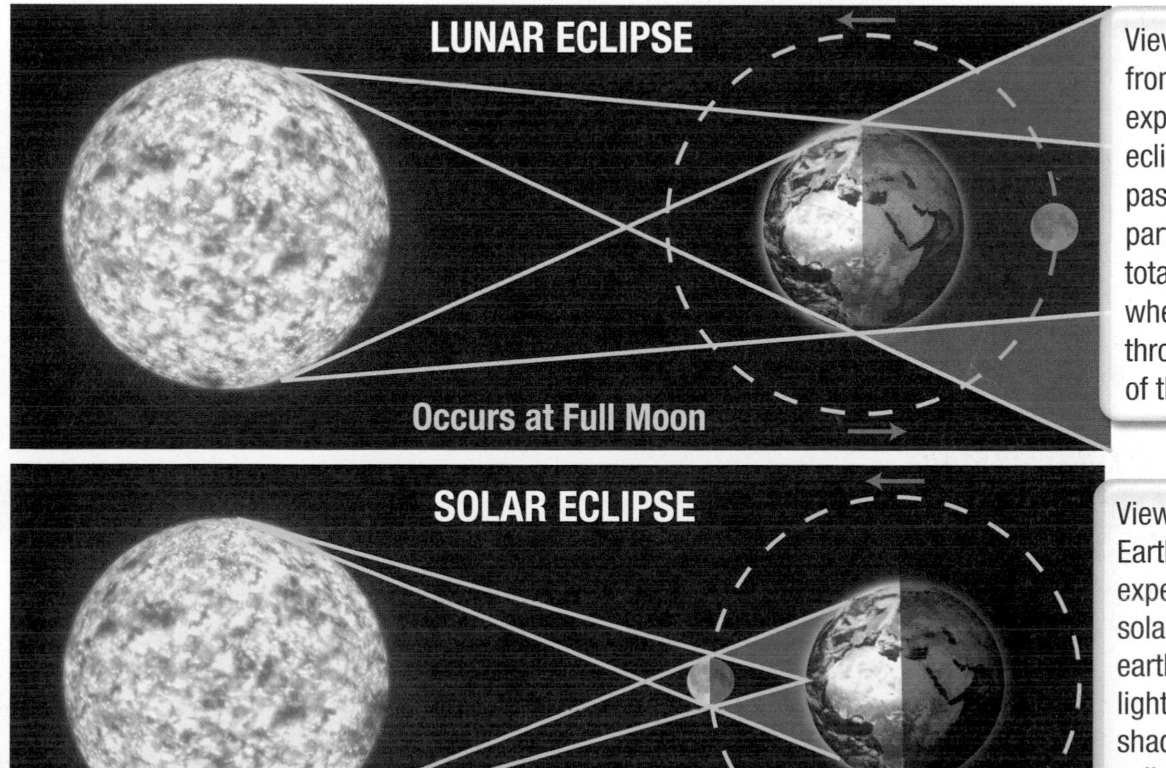

Viewing the moon from Earth, people can experience a partial lunar eclipse when the moon passes though the lighter part of Earth's shadow. A total lunar eclipse occurs when the moon passes through the darker part of the shadow.

Viewing the sun from Earth, people can experience a partial solar eclipse when the earth passes through the lighter part of the moon's shadow. A total solar eclipse occurs when the earth passes through the darker part of the shadow.

TIDES

VOCABULARY

tide ('tīd) the periodic rising and falling of the surface level of ocean water

spring tide ('spring 'tīd) a condition of the greatest difference between low and high tides

neap tide ('nēp 'tīd) a condition of the least difference between low and high tides

Have you ever stayed at a beach all day and wondered what caused the periodic rising and falling of the surface level of the ocean? Tides are mostly the result of the moon's gravitational pull on different parts of the earth. Even though the sun has a greater mass than the moon, the moon is much closer to the earth. This proximity makes the moon's force of gravity the main factor that causes the earth's tides.

Ocean waters continually rise for about six hours and then fall for six hours. High tide occurs on two points of Earth at a time and low tide does the same. The side of Earth that is closest to the moon experiences high tide. Interestingly, the side that is opposite the moon also experiences high tide. Low tide occurs on the sides that are farthest away and between the high tides since the water is being pulled toward the high tides. In about a 25 hr cycle, most locations on Earth's surface experience two high and two low tides.

How does the moon affect the tides at the North and South Poles?

High tide occurs at point A because the moon's gravity is pulling the ocean waters toward the moon more strongly than it is pulling the earth as a whole. High tide also occurs at point B because the earth is being pulled toward the moon faster than the water opposite the moon can keep up with, leaving the water behind. Low tide occurs at points C and D as water flows toward points A and B.

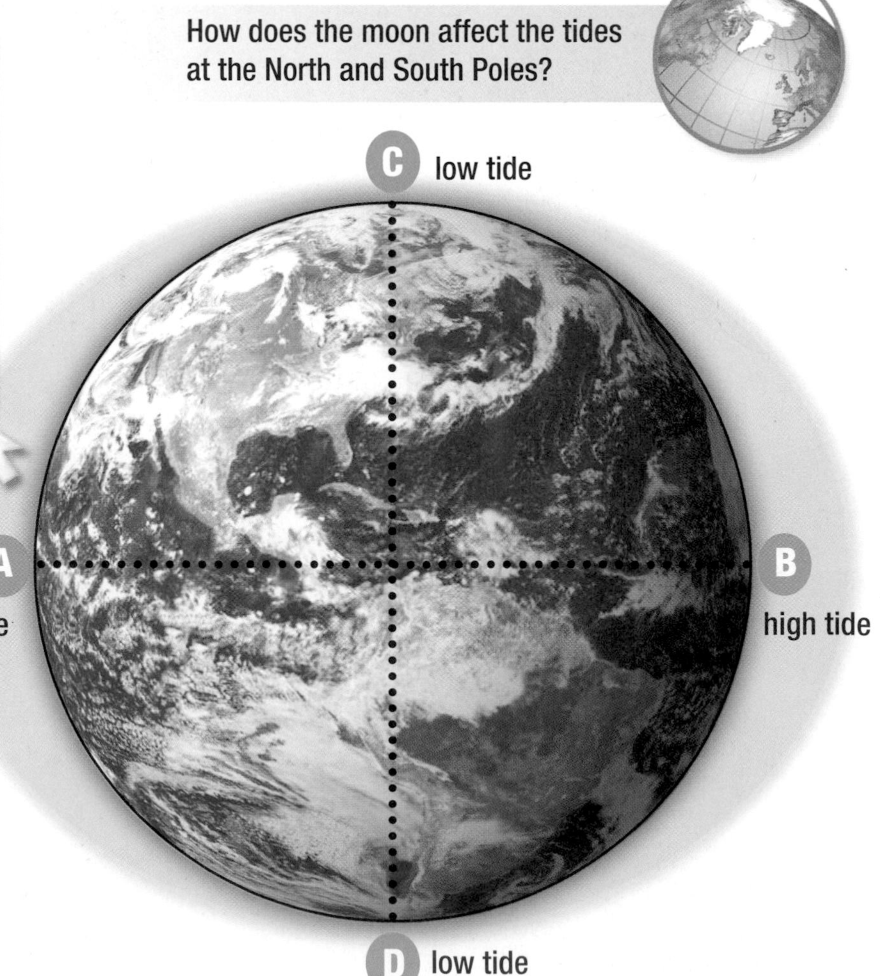

The average distance between the sea levels of low and high tides is 2–3 m (6–10 ft). However, when the sun, the earth, and the moon line up, high tides are higher than normal and low tides are lower than normal. The maximum range between high and low sea levels results from both the sun and the moon exerting a gravitational pull on the earth in the same direction. This condition is called spring tide. Spring tide coincides with the approximate time of full and new moon phases. When an imaginary line between the earth and the moon creates a right angle with an imaginary line between the earth and the sun, neap tide occurs. During neap tide, high tides are lower and low tides are higher than normal because some of the sun's gravity and some of the moon's cancel each other out. The smallest difference between tide levels takes place close to the first and third quarter moon phases. Both spring and neap tides occur twice a month.

A tidal range is the difference in sea levels between high and low tides. The Bay of Fundy in Canada is known for having a tidal range greater than any other place on Earth—up to 17 m (56 ft)! The ocean floor is revealed in this photograph during a low tide.

IN THE FIELD

This inlet of the Rance River, which begins in France, is the location of the largest tidal power plant in the world. The plant harnesses the power of tides and uses water turbines to create electricity. While there are other smaller tidal power plants around the world, the construction cost involved has limited the number of tidal power plants.

SPRING TIDE

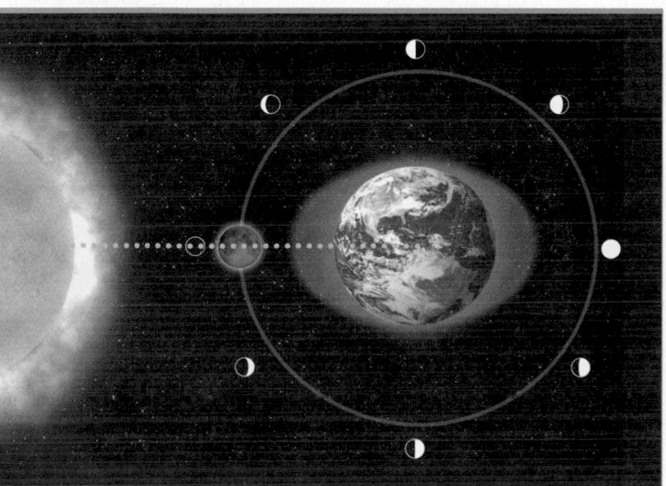

NEAP TIDE

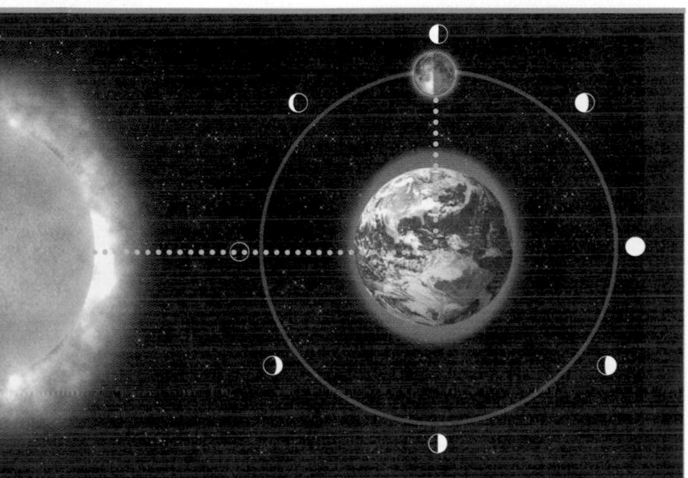

MOON MISSION

VOCABULARY

mare (ˈmär·ā) a dark, flat area on the moon's surface filled with hardened lava

OUTLOOK

Can you imagine standing on the moon and watching the earth rise? Since the moon does not have much of an atmosphere, its sky is always black. How would you feel about living where the sky is in continual darkness? Even in the total darkness of the moon—God is there (Jeremiah 23:23–24).

The 1957 Soviet launch of the *Sputnik 1* satellite triggered the space race. The United States began rapidly working on its own program, sending its first satellite, *Explorer 1*, into orbit in 1958. Later the United States established a government agency to lead the space program, the National Aeronautics and Space Administration (NASA). In 1961, the Soviets sent the first human into space, Yuri Gagarin, in *Vostok 1*. A month later Alan Shepard was the first American in space aboard *Freedom 7*. In 1962, John Glenn became the first American to orbit Earth three times inside *Friendship 7*.

In 1961, President John F. Kennedy announced the Apollo program, an American effort to place a man on the moon by the end of the decade. On July 16, 1969, the United States sent astronauts Neil Armstrong, Buzz Aldrin, and Michael Collins to the moon aboard *Apollo 11*. Four days into their journey, Collins separated the spacecraft into two parts. Collins was in the *Columbia* command module, which continued to orbit the moon. Armstrong and Aldrin were in the lunar module, *Eagle*, which landed on July 20, 1969.

About four days and 384,000 km (240,000 mi) after launch, *Eagle* landed on a flat area on the moon's surface called *the Sea of Tranquility*. Six and a half hours later, Aldrin opened the hatch and Armstrong slowly backed out of the lunar module. In his excitement he stated, "That's one small step for man ...one giant leap for mankind."

The flag on the moon has a pole running across the top of it because there is no wind to blow it. The flag is on the near side of the moon and always faces Earth.

Apollo 11's astronauts learned much about the moon during their 2.5 hour tour of it. However, certain things about the moon, like its bright and dark spots, are visible even without a telescope. The moon's surface includes interesting features. A light-colored high peak on the moon is a mountain called *a highland*. A dark, flat area is a sea called a mare, Latin for *sea*. Lunar seas are not filled with water but with hardened lava. A crater is a large round depression that has been created by the impact of a meteoroid, a rock fragment from an asteroid or comet. Craters exist on the maria and even more so on the highlands.

The moon's characteristics, which cannot sustain life, are quite different than those of the earth. The moon has no atmosphere, wind, or liquid water, and its temperatures vary. When direct sunlight hits the moon's surface, the temperature can rise to 127°C (260°F). Without the sunlight the temperature can reach a frigid –240°C (–400°F).

LINKS

Writing Link
Interview a person that remembers the first moon landing. Have questions such as *What impressed you the most about space as you watched the launch on TV?* prepared in advance. Record the answers to your interview questions and write a paragraph about your interviewee's experience.

History Link
Research other events during the space race. Include missions that the Soviet Union, the United States, and other countries made after *Sputnik 1*. Design a time line with the most important events.

Art Link
Design a poster of the different types of tides or eclipses. Be sure to use creativity, neatness, and color. Include captions for each of your illustrations.

Technology Link
Research what technological failures caused the *Apollo 13* mishap and what technological successes helped bring it back to Earth.

Does the moon change size?

This is a thin sample of rock from the moon, brought back to Earth by the *Apollo 12* mission. Different minerals and their textures are shown in color photographs taken through a microscope using a special light.

The moon's surface has been explored and observed directly for over four decades. *Clementine*, a small spacecraft launched in 1994, found indirect evidence of ice in the Tycho Crater at the moon's southern pole.

UNIT 1

LIFE SCIENCE: CYCLES

UNIT 2

PHYSICAL SCIENCE: TRANSFORMATIONS

UNIT 3

EARTH AND SPACE SCIENCE: PREDICTABILITY

HUMAN BODY: BALANCE

– the study of the body and how its systems work together to keep a person healthy

CHAPTER 13

TRANSITIONS

VOCABULARY

puberty (ˈpyo͞o·bər·tē) the process of physical changes as a child's body develops into an adult

adolescence (a·də·ˈle·səns) the period of time between childhood and adulthood

KEY IDEAS

- Adolescence and puberty involve many physical changes.
- The endocrine system works with the nervous system to stimulate these changes.
- The endocrine system helps to maintain homeostasis, but lifestyle choices also contribute to this balance.

As you get older your body changes from that of a child to that of an adult. The process is called puberty. It occurs somewhere around the age of 12. Girls usually begin puberty earlier than boys.

Adolescence is the time period surrounding puberty. Often with adolescence comes both excitement and concern regarding all the physical changes. Similar to white-water rafters traveling down a river of calm and rough waters, adolescence is like paddling through emotional, intellectual, and social changes. Sometimes the trip is smooth and sometimes it is a bit unsettling. This stage of life requires learning new skills and making wise decisions. This will help you grow in confidence, deal with stress, and interact with others. During this time expect change, plan for it, and remain self-controlled.

QUICK FACT

Changes typical in both boys and girls:

- Height and weight increase.
- Arms/hands and legs/feet grow in size first.
- Chin and nose grow.
- Oil builds up in skin and hair.
- Pimples may appear on face and body.
- Body develops and matures.
- Hair under arms begins to grow.

Changes usually occurring only in girls:

- Body fat increases in abdomen and legs.
- Hips begin to widen.

Changes usually occurring only in boys:

- Muscles develop in the chest and arms with broadening of the shoulders.
- Voice begins to crack and deepen.

Strange but True

Is this alive? Is it found on your body?

THE ENDOCRINE SYSTEM

VOCABULARY

hormone ('hôr·mōn) a chemical messenger that creates a response in the body

gland ('gland) a specialized tissue that produces and releases chemicals

endocrine system ('en·də·krən 'sis·təm) a network of glands that produces and releases hormones

homeostasis (ˌhō·mē·ō·'stā·səs) a state of balance within a cell, organ, or system

pituitary gland (pə·'to͞o·ə·ˌter·ē 'gland) a gland in the brain that stimulates growth and development

A hormone is a chemical messenger that creates a body response, is produced by a gland, and operates like a key. When released into the bloodstream, the hormone key is recognized by certain cells within an organ. These specific cells are the lock into which the key fits. During puberty, hormones stimulate changes in these organs and cause the body to grow and develop.

The endocrine system is a complex group of glands. It communicates by slowly releasing hormones into the bloodstream to remote locations within the body. Hormones trigger body tissues and organs to start, stop, or continue their functions. The nervous system transports electrical signals very rapidly through nerves to and from the spinal cord and brain. These two systems work together to relay messages throughout the body.

The nervous system sends instant messages for quick body responses. The endocrine system regulates slower body processes. Glands release hormones into the blood stream where they are then delivered to specific receiving cells.

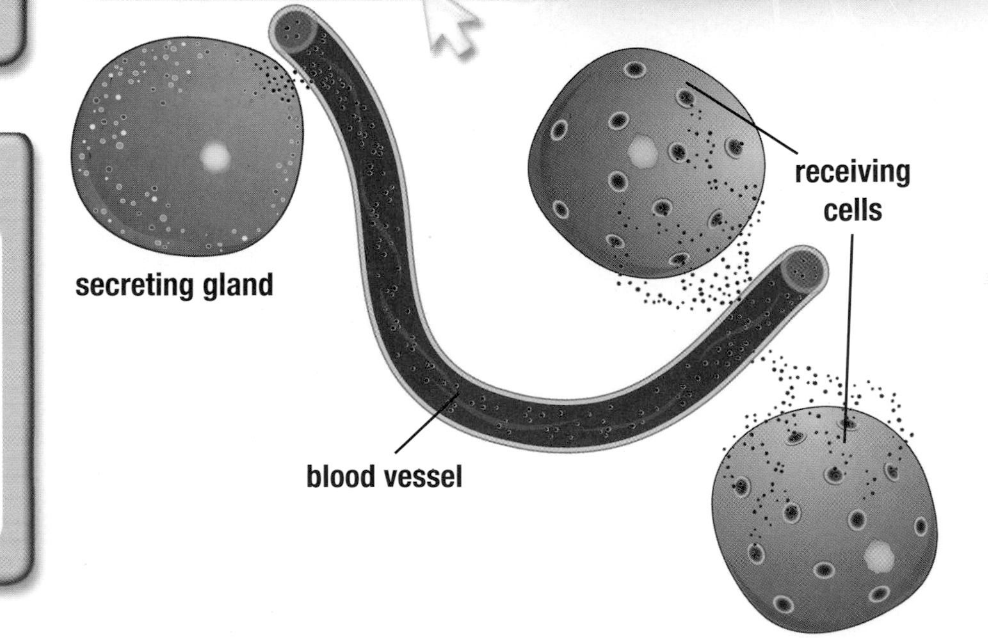

QUICK FACT

The term *hormone* comes from a word that means *to stir up.* During puberty many of these chemical messengers travel throughout a young person's body, stirring up several kinds of changes.

Major parts of the endocrine system include the hypothalamus, pituitary gland, thyroid gland, adrenal glands, pineal gland, and the pancreas. The hypothalamus is the link between the endocrine and nervous systems. It sends various signals to trigger an internal balance within the body called homeostasis. In addition to this, the hypothalamus alerts you about being sleepy or hungry. It also gives the signal for your body to begin puberty.

The pituitary gland is sometimes called *the master gland.* It responds to signals from the hypothalamus to control the other endocrine glands. The pituitary gland also releases the growth hormone. The pineal gland affects the body's sleep cycle. The thyroid gland determines how quickly energy is released from the food you have eaten. The adrenal gland helps you respond to stress and influences the balance of salt and water. These all work together to cause you to grow and develop.

IN THE FIELD

Endocrinologists are doctors who spend up to 10–14 years learning their specialty. First they attend college, next medical school, and then a residency program. Their job is to diagnose and treat illnesses that affect a person's hormone levels. The disorders and diseases of the endocrine system are placed into nine different groups—thyroid, pituitary, bone, reproduction, obesity, growth, hypertension, diabetes, and lipid disorders.

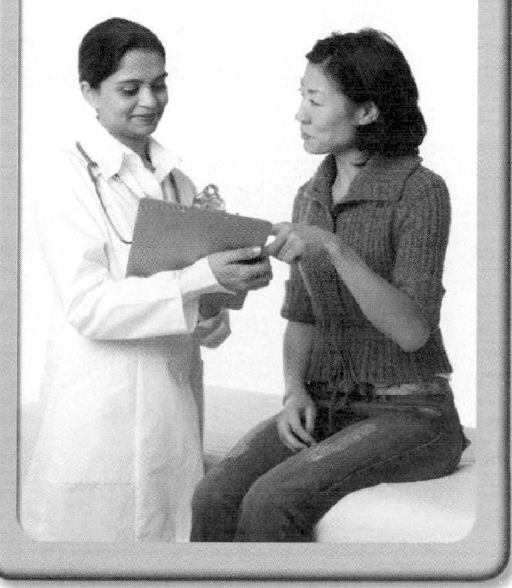

What is adrenaline and what does it do?

hypothalamus

pituitary gland

pineal gland

thyroid gland

adrenal glands

pancreas

SKIN

VOCABULARY

sebum ('sē·bəm) an oil produced by glands located in the skin

acne ('ak·nē) a skin condition that results when mostly excess sebum clogs skin pores

perspiration (pər·spə·'rā·shən) a fluid released by the sweat glands

QUICK FACT

Humans shed tens of thousands of dead skin cells every minute! That adds up to approximately 4 kg (9 lb) per year. A great deal of the dust in houses is made up of human skin cells. Fortunately, the epidermis makes new skin cells to replace the ones that die and flake off.

Hormones affect skin—the largest organ in the human body. In an average size adult, skin covers about 2 m^2 (22 ft^2). It is made of different types of tissue that together perform specific functions. Some of these vital body functions include protection from the external environment; maintaining body temperature; and the sensations of pain, pressure, heat or cold, and touch. Skin cells also produce vitamin D when exposed to sunlight.

The skin's top layer is called *the epidermis*. The epidermis provides a barrier between the underlying tissue and the outside world. Certain epidermal cells are part of the immune system. They prevent harmful bacteria and viruses from entering the body. Interestingly, nails and hair, although made from dead cells, are two types of epidermis. Another layer, called *the dermis*, supports and strengthens the epidermis. It contains many nerves, blood vessels, glands, and hairs.

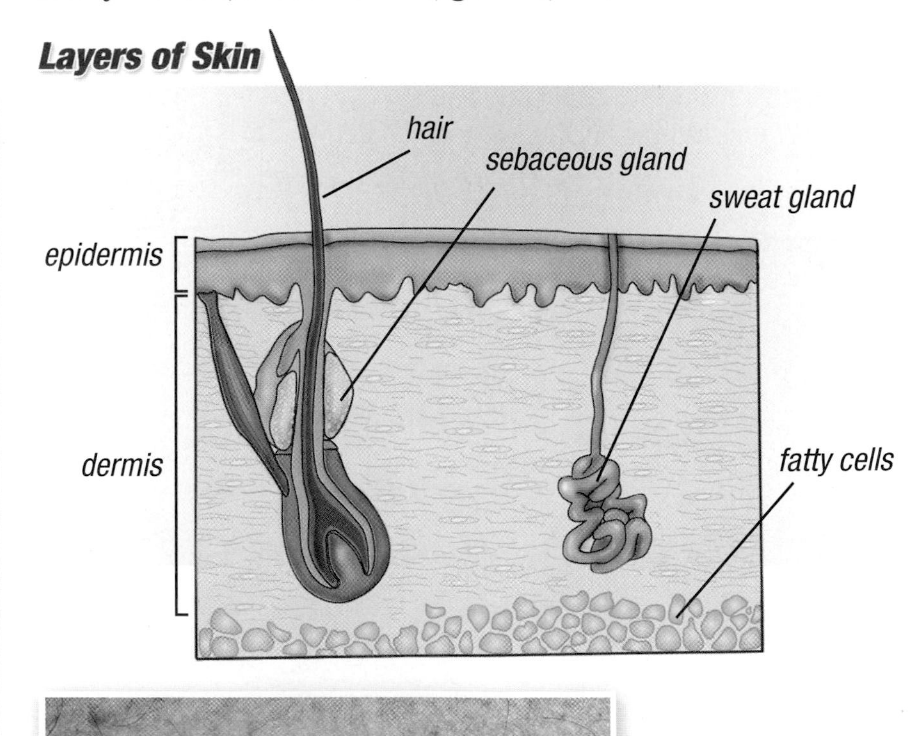

Skin cancer can sometimes be surgically removed. It is important to limit exposure to harmful ultraviolet rays.

The skin has certain glands that release substances directly to the surface of the skin instead of into the bloodstream. Sebaceous glands are located within the dermis and are usually found near each hair strand. These glands release sebum, an oil that travels to the surface of the skin along the hair and through tiny tubes called *ducts*. Sebum waterproofs and protects the skin and hair by preventing dryness. Endocrine glands cause the sebaceous glands to enlarge and produce more sebum during puberty. This increased production can result in acne if pores, openings on the surface of the skin, become clogged with sebum and other substances.

Sweat glands send perspiration to the surface through pores in the epidermis. This watery and salty fluid evaporates, thereby cooling the skin and helping to maintain body temperature. While perspiration itself is odorless, it sometimes reacts with bacteria on the skin to create an odor. Since some sweat glands increase their production during puberty, it is important for adolescents to bathe regularly and use deodorant.

Does a person only sweat when he or she is overheated?

IN THE FIELD

A dermatologist is a physician who diagnoses and treats disorders and diseases of the skin, hair, and nails. He or she must learn about many different fields of science such as microbiology, biochemistry, physiology, and endocrinology. Dermatologists in some countries, such as New Zealand, study for 13 years before they are fully trained in this specialty.

Acne is common in adolescents and includes the appearance of whiteheads, blackheads, pimples, and red lumps, or cysts. Daily skin care may minimize outbreaks and shorten healing time.

Hair grows faster in the summer than in the winter and faster during the day than at night. Beard hairs are the fastest-growing human hairs. If a man never shaved his face, his beard could become 9 m (30 ft) long in his lifetime.

VOCABULARY

long bone ('long 'bōn) one of several elongated parts of the human skeleton

QUICK FACT

Although human fingers have bones, they do not have any muscles. Instead, they transfer force from the muscles of the hand, wrist, and arm to make any of 58 possible movements. When a concert pianist performs, extensive coordination of 35 muscles in each hand must occur.

Have you ever experienced rapid growth? Perhaps you have known someone who grew several centimeters (inches) in one year. Major growth spurts occur during puberty because the hypothalamus sends messages to the pituitary gland. The pituitary gland then sends growth hormones into the bloodstream. These chemical messengers stimulate the growth of bones, muscles, and internal organs. In addition to the endocrine system, genetics, nutrition, sleep, and exercise all affect the growth rates, height, and weight of adolescents.

Doctors can gain insight into a patient's growth pattern by evaluating the person's bones. Long bones make up the elongated parts of the skeleton. Arms and legs each have three long bones. The thigh bone, or femur, is the longest bone in the human body. Each long bone includes growth plates, which are locations where the lengthening of the bone occurs.

At what time of day are the most growth hormones produced?

Each long bone has a growth plate at each end, as shown here at the knee in blue. The growth plate affects the length and shape of the mature bone. Once growth stops, the growth plate hardens into solid bone.

Adolescents can experience broken bones, or fractures. Sometimes bones are stronger after they have healed from breaks.

Adolescents can grow approximately 5–10 cm (2–4 in.) every year of a growth spurt! The rate at which growth occurs varies greatly. In general, girls experience most of their adolescent growth between the ages of 10 and 12, and many reach their full height by age 14. Boys typically grow rapidly between the ages of 12 and 14, and continue getting taller until about age 17.

Your adult height and weight will probably be similar to one or both of your parents. However, besides hormonal and genetic factors, lifestyle choices also affect growth. For instance, bones need calcium to develop properly. If there is not enough of this mineral in the bloodstream, certain cells will break down bone to obtain it. Exercise can increase muscle mass, causing these cells to create thicker bones to support the muscle. Healthy food choices and regular exercise are important during puberty and throughout life.

OUTLOOK

The genetic material in the DNA molecule is one of the main factors that determines a person's height and weight. As a result, adolescents often resemble one or both of their parents in these traits. However, even identical twins, who have almost the same DNA, will develop unique fingerprints. Could this be part of what God had in mind when He caused the Psalmist to write in Psalm 139:13–14, "You knit me together in my mother's womb. I praise you because I am fearfully and wonderfully made"?

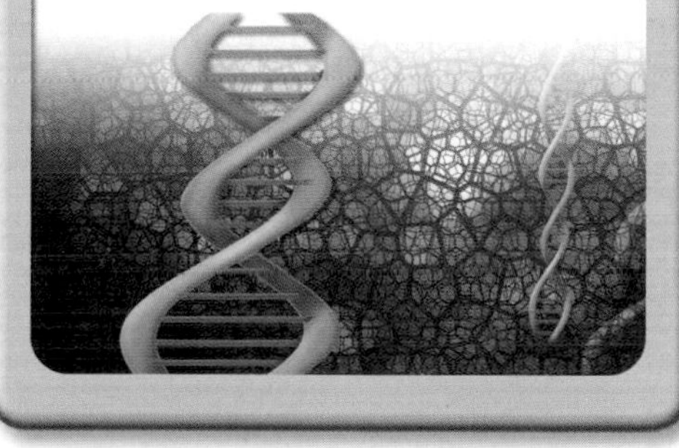

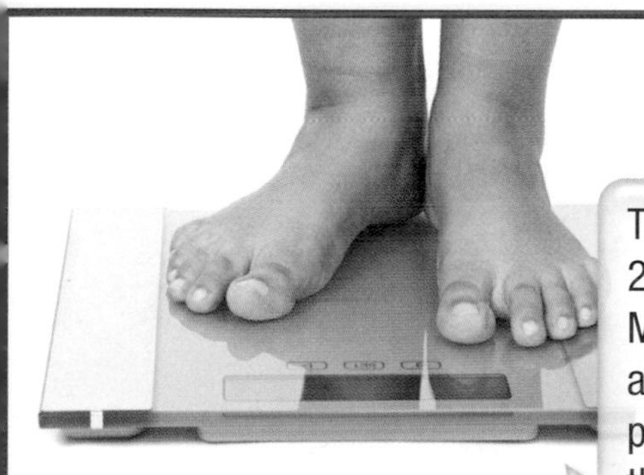

The condition of being more than 20% overweight is called *obesity*. Maintaining a healthy weight helps to avoid diabetes—a disease in which patients have high levels of sugar in their blood.

Doctors refer to recommended weight charts. They emphasize nutrition and exercise as ways to achieve a healthy weight.

Strange but True

This is a microscopic image of an eyelash. Live hair cells develop from hair bulbs located in the dermis. However, as hair grows, it changes into a hard protein called *keratin*. During this process the cells die.

TEETH

VOCABULARY

wisdom teeth ('wiz·dəm 'tēth) the third molars, usually the last teeth to appear

cavity ('ka·və·tē) an area of tooth decay caused by prolonged exposure to bacteria

QUICK FACT

There are 53 facial muscles. It takes more muscles to form a frown than a simple smile. Five pairs of facial muscles play the largest role in smiling. Almost all of the muscles are involved in an exaggerated smile. Smiling and laughing have been scientifically proven to stimulate the immune system. These activities help restore homeostasis to the body by keeping the amount of a hormone called *cortisol* at a proper level. So, smile and laugh—it is good for you!

Just as your body goes through a transitional time, so do your teeth. Teeth are part of the skeletal system as well as the digestive system. Humans possess two sets of teeth during their lifetimes. The first to develop are the primary teeth. The permanent, or secondary, teeth replace the primary ones that fall out. This usually happens between the ages of 6 and 14. The last to break through the gums are typically the third molars, or wisdom teeth. They often appear during or after late adolescence—between 17 and 21 years of age.

A sticky coating on teeth is known as *plaque* and forms when food particles mix with bacteria. Bacteria digest the carbohydrates in food. This chemical process produces acids, which often dissolve tooth enamel and form cavities. If not treated, the decay can reach the dentin layer of the tooth. Fortunately, saliva helps protect the exposed tooth surfaces.

Why are the third molars called *wisdom teeth*?

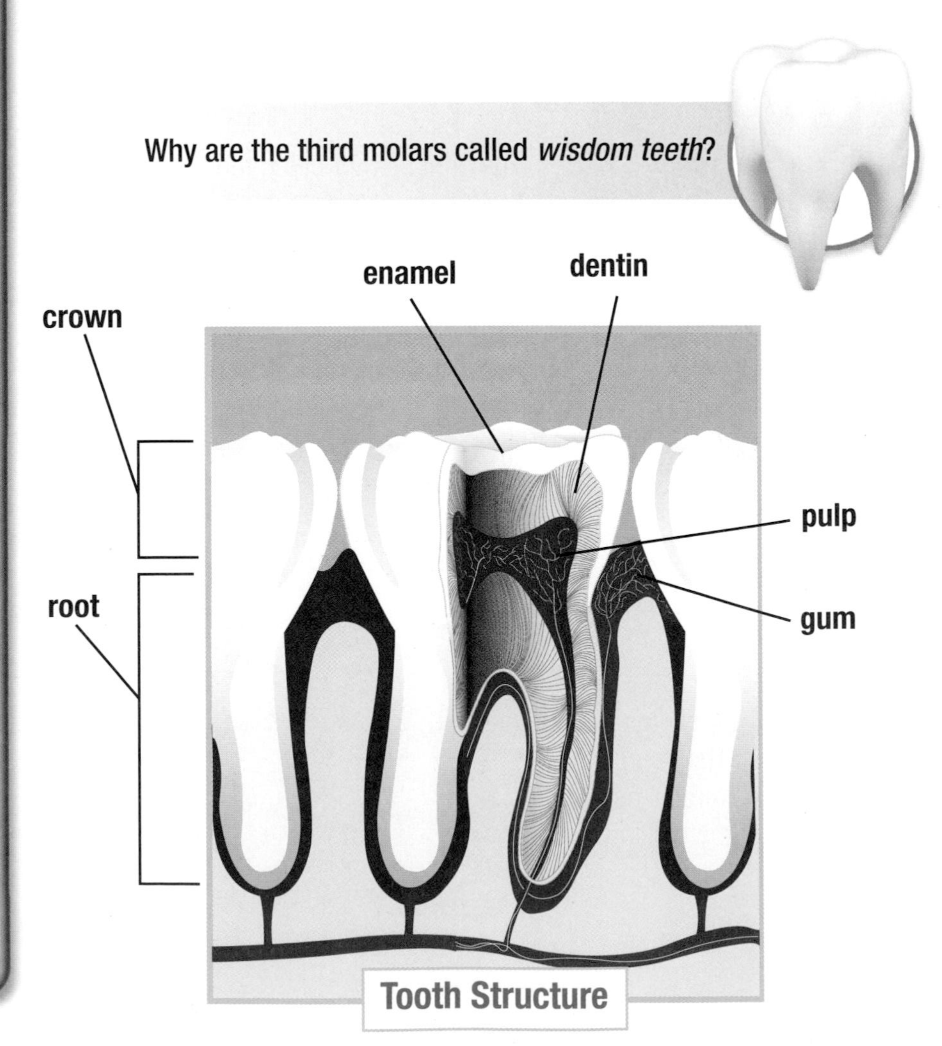

Tooth Structure

Bacteria in the mouth can also cause bad breath. The scientific word for bad breath is *halitosis*. Brushing at least two times a day and flossing nightly often cures this odor. Many toothpastes and treated water contain fluoride, a mineral that helps make teeth strong and prevents cavities from forming.

Teeth serve several purposes. They allow you to chew and speak. They also greatly affect your appearance by adding shape and form to your face. Taking care of your teeth during adolescence may prevent serious problems later in life. It is important to see a dentist regularly for a thorough checkup. In addition, avoiding too many sugary snacks or starchy foods and brushing after meals is important. Teeth can also be knocked out accidentally. It is wise to wear a mouth guard when playing sports.

IN THE FIELD

Orthodontists are dental specialists who receive two more years of training after four years of dental school. Orthodontics is the branch of dentistry involving the correction of teeth and jaw alignment through the use of braces. Crooked teeth are more difficult to keep clean and, therefore, more likely to develop cavities. They may also interfere with proper chewing and facial development.

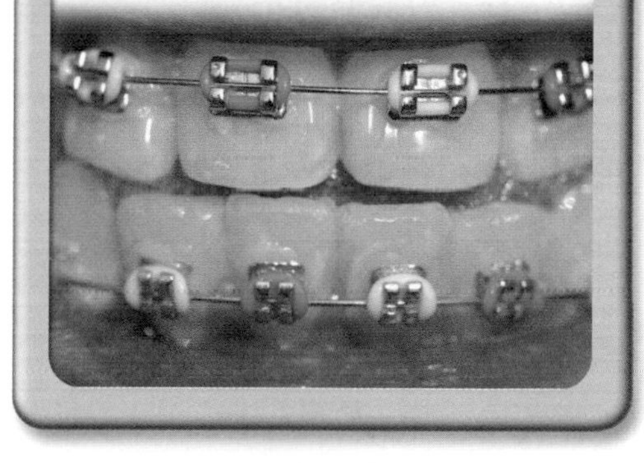

Dental hygienists clean, floss, and polish teeth. Since you only have one set of permanent teeth, you should take good care of them.

A knocked-out permanent tooth is a dental emergency. The tooth should be picked up by the crown and then placed in a cup of milk or back in the person's mouth. Going to a dentist or hospital immediately is crucial. A tooth has the best chance of surviving if it can be returned to its socket within 30 min after coming out.

BODY CLOCK

IN THE FIELD

Modern sleep/wake research is called *chronobiology*. Scientists use electroencephalograms (EEGs) to measure a person's electrical brain waves while he or she is sleeping. Small electrodes from the machine are attached to a patient's head to record his or her brain waves. A relaxed, alert person will experience about 10 waves per second while a person in the deepest stage of sleep will only have about 4 per second.

The human body has an internal clock that controls its daily rhythms, including the sleep cycle. The endocrine and nervous systems work together to affect sleep. At night when it is dark, the pineal gland begins production of a hormone that induces sleep. Just the opposite happens during daylight hours. When light reaches the eye's retina, the optic nerve sends a signal to the hypothalamus. The hypothalamus then triggers the pineal gland to stop producing the hormone.

The pineal gland is a pinecone-shaped structure in the brain. In an adult it is about 8 mm ($\frac{1}{3}$ in.) long and weighs approximately 0.1 g (much less than 0.1 oz). This gland is larger in children. With the onset of puberty, it begins to shrink and produce less of the sleep hormone than before. As a result adolescents can experience difficulty going to sleep or waking up early. The pineal gland affects the body's development during its sleep and wake cycles.

What does this toddler have in common with a teenager? They both need more sleep than at any other time in their lives. Toddlers and preschoolers need about 10–12 hours of sleep per day. Adolescents need approximately 9–10.

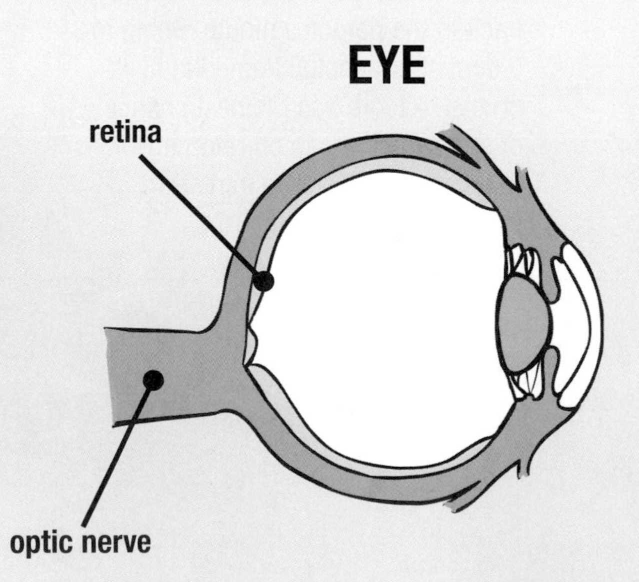

A disruption of normal sleep patterns can generate problems. Examples include mood changes, abnormal vision and speech, difficulty thinking clearly, fatigue, or confusion. A person who is deprived of sleep might even cause accidents. For example, the Exxon Valdez oil spill was partially due to the fatigue of night shift workers.

There are two kinds of sleep—nonrapid eye movement (NREM) and rapid eye movement (REM). Together, they consist of five stages. The first four stages are NREM sleep. They begin with drowsiness and gradually progress toward a deep sleep. Brain waves, breathing, and heart rates slow and body temperature and blood pressure drop. The fifth stage is REM sleep. During REM the large muscles completely relax but the body's functions and activities tend to speed up. This stage is also characterized by rapid eye movements and intense dreaming. The amount of REM sleep is highest during infancy and decreases during adolescence.

Some scientific research demonstrates a link between adequate sleep and successful learning.

This is a graphic record of what happens during REM sleep. The red box highlights an EEG, or the brain waves of an individual. The red line just below the red box represents the rapid eye movements.

LINKS

Technology Link
Use the Internet to learn how light therapy is given to people with sleep disorders or seasonal affective disorder. Print out several images of different types of phototherapy lamps. Combine the pictures into a booklet that includes a title page and a two-paragraph summary of light therapy.

History Link
Write a report about sideshows in the Barnum and Bailey circuses of nineteenth century America. Highlight the ones that featured people who may have had growth hormone disorders.

Language and Art Link
Choose one of the following activities: (1) Draw a comic strip about adolescence that includes speech bubbles, (2) Write an acrostic using the word *puberty*, (3) Create a poem about transitions, or (4) Write a tongue twister on the topic of homeostasis.

CHAPTER 14

DISEASE

Disease occurs when there is a breakdown in the normal function of an organism. For many years people did not understand the causes of disease. Many mistakenly thought that an evil spirit, a curse, or bad luck was the reason. Cleanliness was not considered a factor. Not until the 1800s was it discovered that many diseases are caused by microorganisms.

Today, diseases are classified into different categories in an effort to help in their diagnosis and treatment. A diagnosis is made by health care professionals after a thorough examination and observation of the symptoms the patient is showing. A proper treatment plan is then prescribed, which sometimes includes medicines.

VOCABULARY

disease (di·'zēz) any abnormal condition that causes illness in a living organism

KEY IDEAS

- Diseases can be classified as *infectious* or *noninfectious*.
- Infectious diseases are caused by pathogens.
- Genetic, social, and environmental factors affect disease.
- Body systems work together to fight disease.

IN THE FIELD

The World Health Organization (WHO) is a special agency developed by the United Nations. It employs over 8,000 people from more than 150 countries. Its main mission is to combat disease and to promote the general health of people worldwide. This important organization works with the governments of countries to urge the use of programs that prevent and treat disease. It also educates and informs people about the quality and safety of their water, food, air, and medicines.

The human body serves as a host for many microorganisms. Pictured here are three types of microbes found in the mouth. The round bacteria help to break down sugar but are also responsible for tooth decay. The rod-shaped bacteria cause diseases such as gingivitis, or gum disease. The two larger cells are a yeastlike fungus, which can cause thrush, a yeast infection of the mouth. With all these microorganisms in the mouth, it is important to regularly brush your teeth!

Why do feet sometimes have a bad odor?

Strange but True

PATHOGENS

VOCABULARY

pathogen (ˈpa·thə·jən) an organism or particle that causes disease

infectious disease (ˈin·fekt·shəs di·ˈzēz) a contagious disease caused by pathogens

noninfectious disease (nän·ˈin·fekt·shəs di·ˈzēz) a disease that is not caused by pathogens and is not contagious

Different diseases have different causes. Some are a result of pathogens, which are organisms or particles that lead to disease. Pathogens are often microorganisms that can only be seen with a microscope. A disease that originates from a pathogen is called an infectious disease. These diseases are usually considered contagious because they can spread from one individual to another. On the other hand, a noninfectious disease is not caused by pathogens and cannot be spread from person to person.

Pathogens are classified into groups. Four common groups include bacteria, viruses, fungi, and protists. Bacteria are unicellular and cause a wide variety of diseases. Leprosy and conjunctivitis, or pink eye, are two examples of bacterial infections. Viruses are tiny particles much smaller than bacteria. They are responsible for hundreds of kinds of colds and flu, as well as other diseases. Fungi, which include molds and yeasts, like to grow in warm, moist places. Athlete's foot is a common fungal disease. Protists are also unicellular organisms. Many protists that cause disease live in unpurified water and ingesting them can lead to serious illness.

QUICK FACT

Chicken pox is a common childhood disease. It is characterized by pox, or small sores on the body. The infectious pathogen is the *Varicella zoster* virus. A vaccine against chicken pox was developed in 1974 and can prevent this disease. It is recommended that young children receive this vaccine. Cases of chicken pox range from mild to severe.

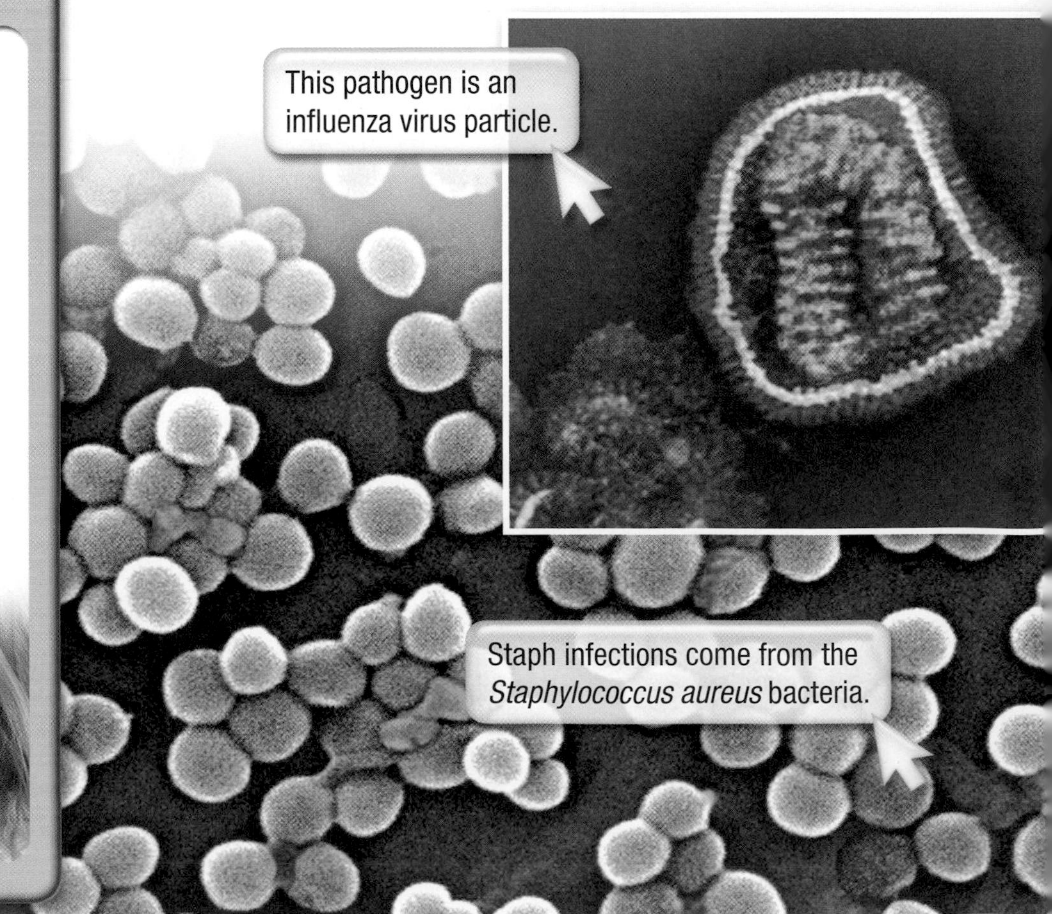

This pathogen is an influenza virus particle.

Staph infections come from the *Staphylococcus aureus* bacteria.

Infectious diseases are contagious because of the pathogens that cause them. These organisms can be spread in a number of ways. Pathogens commonly pass from one person to another. This can happen directly, as in a hand shake, or it can happen indirectly when someone coughs or sneezes. Some pathogens live naturally in soil, food, or water. Botulism, or food poisoning, is caused by bacteria that live in the soil. If people eat food or drink water that is contaminated with these organisms, they may become sick.

Animals can be carriers of pathogens too. If an infected animal bites someone, the pathogens can pass to the individual. Mosquitoes and ticks are two very common carriers of pathogens. Also, people can become infected with pathogens by coming in contact with contaminated objects. This is how colds and flu are often spread. If an infected person touches something like a drinking glass or door knob, the pathogens can be passed to anyone who also uses that object. For this reason, proper hygiene, a clean environment, and hand washing are very helpful in reducing infection and the spread of disease.

OUTLOOK

In His infinite wisdom, God provided practical solutions for the prevention of the spread of infectious diseases. Well before people knew what caused disease, God gave specific laws in the Old Testament that taught not only obedience, but were also a means of protection. Leviticus includes the first record of quarantine, or the isolation of people with infectious diseases. This effectively contained and prevented the spread of the disease. In Numbers the priests were instructed to use hyssop soap and water to clean their hands and clothes after touching the blood of sacrificial animals. In Deuteronomy, the Israelites were instructed to keep their waste out of the camp, which reduced water-borne diseases. How marvelous it is to know that God's laws protected people from infections, even when people were unaware of the diseases.

What does this moldy loaf of bread have to do with fighting against infection?

Trichophyton rubrum is a type of fungus that causes hair, skin, and nail infections.

This parasitic amoeba is a protist. It is a pathogen that results in diarrhea and ulcers.

INFECTIOUS DISEASES

VOCABULARY

acute (ə·ˈkyo͞ot) beginning suddenly and lasting for a short time

chronic (ˈkrä·nik) lasting for a long time and/or recurring frequently

virus (ˈvī·rəs) a tiny, nonliving particle that can infect cells and cause disease

QUICK FACT

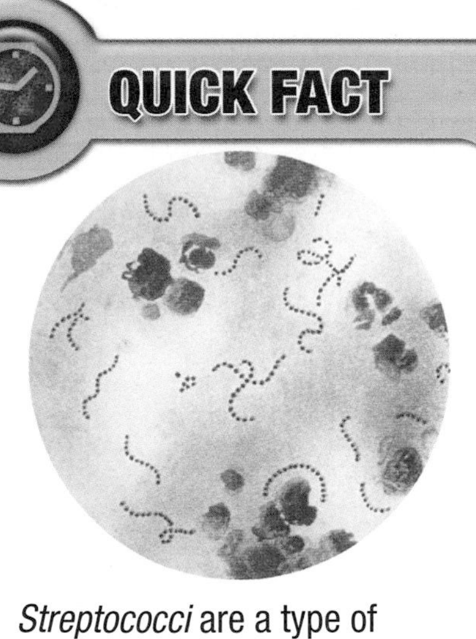

Streptococci are a type of bacteria that are round and grow in pairs or chains. Some cause strep throat, rheumatic fever, and scarlet fever. Others bring about pneumonia and meningitis. Not all of them are pathogens. For example, one kind is used in the manufacture of certain cheeses and yogurts.

Pathogens make you ill by damaging cells in your body. An illness that comes on suddenly or is short in duration is considered an acute disease. Chronic, on the other hand, refers to a disease that lasts a long time. Bacteria are often the cause of acute infectious diseases. Some bacteria harm body cells directly. Other bacteria produce poisons that damage cells indirectly. Common bacterial diseases are bladder, kidney, and ear infections, tonsillitis, tuberculosis, tetanus, whooping cough, and botulism, or food poisoning.

Bacteria are masters of change. This is a serious health problem because they can adapt to antibiotics, such as penicillin, that are used to fight against them. Doctors warn that one of the biggest triggers of resistance to these drugs happens when patients do not finish their entire prescription. Sometimes patients do this because they feel better or because the antibiotics are costly. Not taking all the medicine prevents the bacteria from being completely destroyed. This allows them time to adapt and change, which can make them even more dangerous.

Symptoms of strep throat can include red throat, fever, and yellow pus-filled membranes over the tonsils.

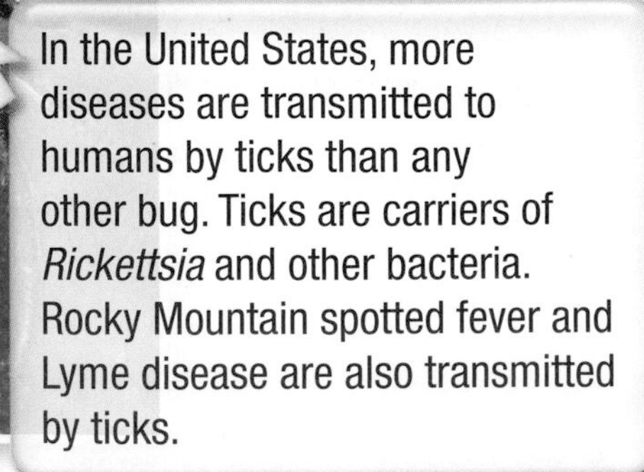

In the United States, more diseases are transmitted to humans by ticks than any other bug. Ticks are carriers of *Rickettsia* and other bacteria. Rocky Mountain spotted fever and Lyme disease are also transmitted by ticks.

A virus is a tiny particle that contains genetic material surrounded by a protein coat. Scientists do not consider viruses to be living because they are not made of cells, and they cannot reproduce outside of their host's body. In order to reproduce, a virus attaches itself to a cell. It then injects its genetic material into the cell. This overrides normal cell functioning and causes the cell to produce copies of the virus. The cell then releases the viruses and the cycle begins again.

Viruses are several hundred times smaller than cells. The size and shape of each virus is different. They can invade and cause a variety of infectious diseases in all forms of life. The tobacco mosaic virus is a well-known plant virus. Rabies is a viral disease that infects both animals and humans. Some common human viral infections are measles, mumps, hepatitis, and yellow fever.

What are the smallest pathogens?

Viruses can be shaped like rods, spirals, spheres, and threads. Bacteriophages, viruses that infect bacteria, are often shaped like tiny spaceships.

IN THE FIELD

Technology plays a major role in the study of disease. Access to computers and the Internet allows for rapid sharing of information about the location and cause of diseases worldwide. Additionally, technology allows for a way of circulating information. This enables medical professionals to stay informed and educated about recent research and advancements in science and medicine. Health care professionals can then educate the people who live in their rural communities about sanitation, health care, immunizations, and medications. All of this helps keep diseases from spreading.

Bacteria become attracted to the sweat on feet and feed on it! The bacteria then excrete substances that have strong odors that make feet smell.

NONINFECTIOUS DISEASES

VOCABULARY

genetic disorder (jə-ˈne-tik dis-ˈôr-dər) an abnormal condition that is usually inherited through genes

allergy (ˈa-lər-jē) an abnormal reaction to a foreign substance

allergen (ˈa-lər-jən) a substance that provokes allergies

Noninfectious diseases are not caused by pathogens and are not contagious. Many are chronic, such as cancer, diabetes, heart disease, stroke, asthma, and arthritis. Some are acute, such as strokes and appendicitis. Because of technology and advancements in science, infectious diseases occur less often. Noninfectious diseases have become more common than they used to be.

Genetic disorders are noninfectious diseases that are usually inherited. They are caused by either mutations in the DNA of genes or by changes in the structure or number of chromosomes. Common genetic disorders are Down syndrome, hemophilia, sickle cell anemia, and color blindness. Down syndrome is an example of a change in the number of chromosomes. Individuals with this disorder have an extra copy of a certain chromosome. Color blindness, hemophilia, and sickle cell anemia are examples of gene mutations. Family history is important when discussing genetic disorders because several common diseases tend to be passed down in families from one generation to the next. These common diseases usually result from an interaction between environmental factors and genes.

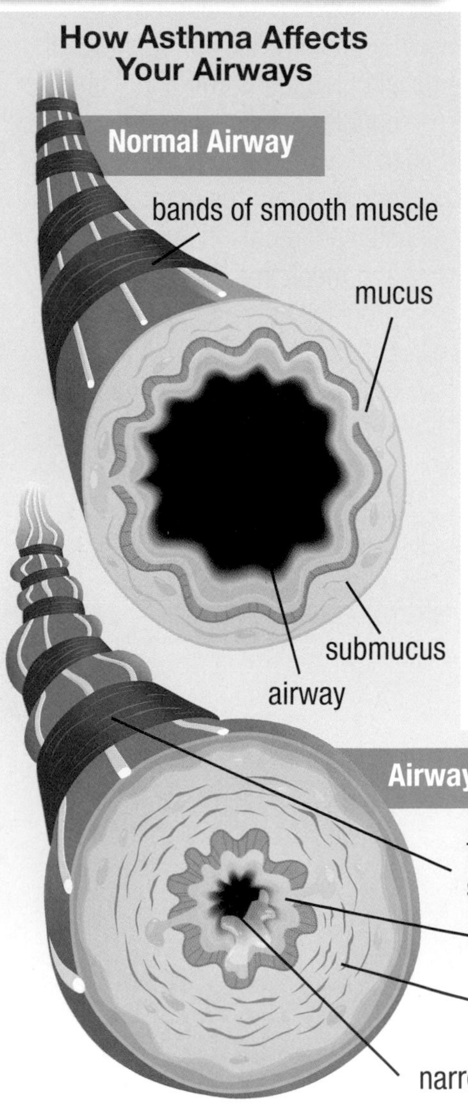

Arthritis causes the inflammation and swelling of joints. There are over 100 types of arthritis. Symptoms include pain, redness, and loss of motion in the joints. It is not yet known what causes this noninfectious disease that affects millions of people.

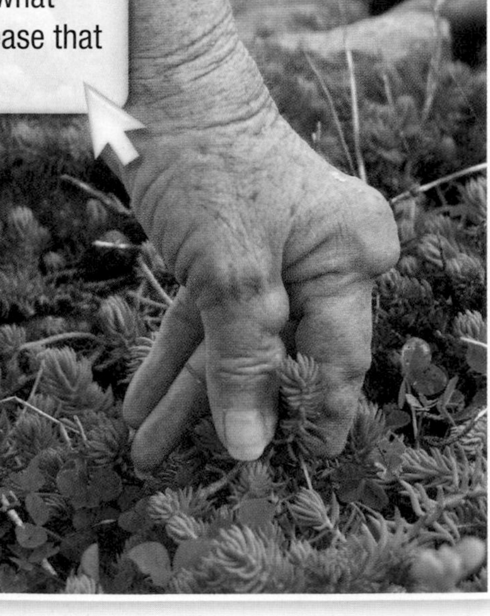

An allergy is a noninfectious disease produced by the immune system's irregular reaction to a substance in the environment. When you inhale, touch, or swallow a foreign substance, special white blood cells are responsible for identifying it. If the foreign substance is a pathogen, the white blood cells trigger your immune system to destroy it. If it is harmless, nothing happens. Sometimes the white blood cells mistake a harmless foreign substance, such as pollen, for a harmful object. The pollen then becomes an allergen. The immune system responds by attacking the particle, causing an allergic reaction. Pollen, dust, mold, certain foods or medicines, and animal dander are all different types of allergens. Poison ivy is a plant that is also an allergen for many people. Allergies can range from mild to very severe.

QUICK FACT

One of the most common acute diseases to occur in children and young adults is appendicitis. The appendix is a small structure attached to the large intestine. It acts as a storehouse for the beneficial bacteria that reside in the intestines. Appendicitis is an inflammation of the appendix. Sometimes matter blocks the opening to the appendix. Bacteria begin to invade the wall of the appendix and the body reacts to this by attacking the bacteria. Symptoms can include sudden sharp abdominal pain, spasms, tenderness, nausea, and vomiting. The appendix can also rupture, releasing its infectious contents into the abdominal cavity. Though appendicitis is caused by bacteria, it is not a contagious disease.

This image magnifies pollen from a variety of plants about 500 times. While pollen is not an allergen to everyone, the white blood cells sometimes mistake them for pathogens.

Dust mites are found everywhere that dust is found. They also can be found on fingernails, hair, pollen, fungi, bacteria, and animal dander. Some people have allergic reactions not to the dust mites themselves, but to the proteins in their feces.

THE IMMUNE SYSTEM

VOCABULARY

lymphatic system (lim·'fa·tik 'sis·təm) a major component of the immune system that protects the body from pathogens

lymph node ('limf 'nōd) a rounded mass of tissue that filters fluids from other body tissues and traps pathogens

immune response (i·'myo͞on ri·'späns) the series of reactions that take place when white blood cells fight against infection

The human body has the amazing ability to defend itself from the invasion of pathogens. The immune system includes barriers such as the skin, mouth, stomach, and breathing passages that prevent harmful organisms from entering. Despite these barriers, some pathogens still get inside the body and damage cells. When this happens, the lymphatic system carries the foreign invaders away from the body's tissues and blood.

The lymphatic system consists of a network of vessels similar to veins. These lymphatic vessels carry lymph, a watery fluid that contains dissolved materials and special white blood cells. As lymph flows through the vessels, it passes through lymph nodes, which trap pathogens. The white blood cells in the lymph begin to destroy the invaders. Lymph nodes often become swollen because the white blood cells are multiplying in order to fight the infection.

Some other components in the immune system are the thymus, the spleen, and the tonsils. The thymus gland changes some white blood cells into a different type of white blood cell. The spleen filters the blood, and the tonsils trap pathogens and fight infection.

What are adenoids and what purpose do they serve?

The spleen filters blood in the same way that the lymph nodes filter lymph. White blood cells in the spleen react to pathogens and set about to destroy them. Other white blood cells engulf the pathogens, as well as any damaged cells and old red blood cells. The spleen is also responsible for maturing red blood cells and certain white blood cells.

The lymphatic system works with both the immune system and the cardiovascular system to protect the body from disease. Lymph nodes are located throughout the body but are concentrated in the neck, armpits, and groin areas.

Another defense against pathogens is the immune response. This is a series of steps in which white blood cells located in the blood and lymph identify and destroy foreign invaders. There are two main types of white blood cells. The first type engulfs the invaders. The second type is responsible for identifying and distinguishing one pathogen from another. They can also produce proteins called *antibodies* that mark the pathogen for destruction. Some antibodies make the pathogens clump together so they can be surrounded easily. Other antibodies keep the harmful organisms from attaching to body cells.

All white blood cells are first produced in the bone marrow. Some stay there to mature. Some travel to the thymus to develop, and others move to the spleen and mature.

Sometimes infections cause a fever. When the white blood cells multiply to fight the infection, they begin working faster to destroy the invaders. This causes the body to heat up, increasing the body temperature. While a fever may not feel good, it means your body is working to defend itself.

IN THE FIELD

Tonsillectomies are surgical procedures that doctors perform to remove tonsils. This may be suggested when someone suffers from frequent occurrences of acute tonsillitis or from chronic tonsillitis. Strep throat infections that take place repeatedly might require a tonsillectomy. Difficulty eating, swallowing, and breathing because of enlarged tonsils are also reasons for this procedure. Otolaryngologists are medical doctors who often perform tonsillectomies. These doctors specialize in the diagnosis and treatment of ear, nose, throat, head, and neck disorders. Removing the tonsils is one of the oldest surgical procedures performed on humans, dating back to at least 30 AD.

What are leukocytes?

The monocyte shown in blue is the type of white blood cell that engulfs pathogens and other cells.

T cells and B cells are types of white blood cells. T cells detect the presence of pathogens. B cells make antibodies. T cells develop in the thymus. The larger B cells are produced and mature in the bone marrow. Shown here is a T cell in yellow and a B cell in green.

IMMUNITY

VOCABULARY

immunity (i·ˈmyo͞o·nə·tē) the body's ability to protect itself from pathogens before they cause disease

active immunity (ˈak·tiv i·ˈmyo͞o·nə·tē) a type of immunity in which the immune system produces antibodies in response to the presence of a pathogen

When a pathogen invades a body, the immune system produces antibodies that help destroy the pathogens. This process provides the body with immunity. The special white blood cells that manufacture antibodies respond by producing a specific antibody for that pathogen. For example, when the chicken pox virus enters your body, the antibodies that are produced are only effective on that particular virus. They cannot destroy any other kind of virus. Your white blood cells have the ability to make over one billion different kinds of antibodies.

When an antibody is made, certain white blood cells called *memory cells* are programmed to remember both the invader and how to make the antibody. The next time the invader enters the body, the memory cells trigger the production of those specific antibodies so quickly that the person usually does not become sick. This type of reaction by the immune response builds active immunity.

QUICK FACT

Even though your body has the ability to develop immunity to certain diseases, no one is immune to all diseases. Some diseases you may only get once, such as chicken pox or measles. Other diseases, like strep throat, can keep recurring each time the pathogen enters your body. Antibiotics are chemicals that kill or slow the growth of a pathogen. They can help fight recurring diseases that are caused by bacteria or some fungi. Antibiotics are not effective against viruses.

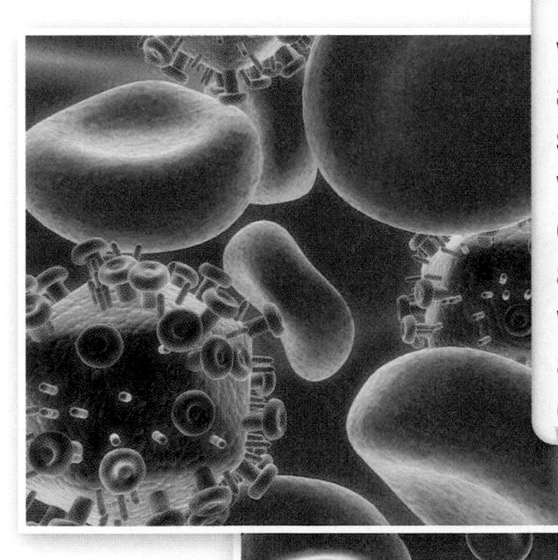

The human immunodeficiency virus (HIV) is a type of virus that attacks the immune system. It specifically targets the T cells, which are the white blood cells that recognize pathogens and send signals to other white blood cells to produce antibodies, preventing these cells from functioning.

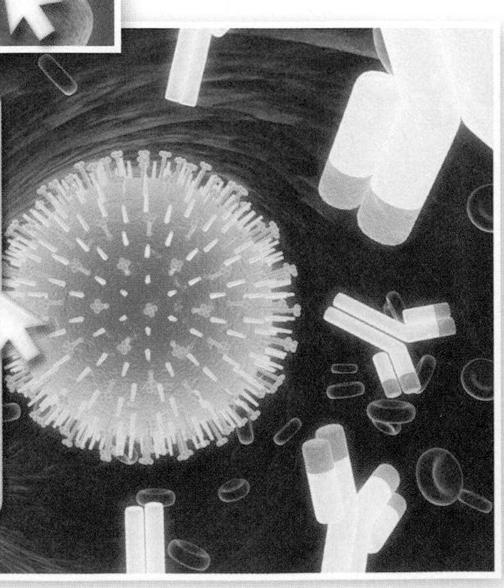

The first time this flu virus enters the body, it takes about two weeks for the white blood cells to make effective antibodies. That is why you get sick the first time. The next time, however, the memory cells recognize the pathogen and the antibodies are made in just a few days.

Your body acquires active immunity in two ways. One is by getting the disease first. The other is from vaccinations. When you receive a vaccination or immunization, dead or weakened pathogens are introduced into your body. This triggers the immune system to develop antibodies and memory cells to fight the disease if you are ever exposed to it. Some vaccinations, such as the mumps vaccine, provide immunity for a lifetime. Some can only provide immunity for a specific number of years, such as the tetanus vaccine. In this case, booster shots must be received to keep the body protected. Vaccinations can be administered by injection, oral drops, or nasal spray.

Sometimes a person can acquire passive immunity. This happens when antibodies come from a source other than the person's body. A baby receives passive immunity from its mother before birth. The antibodies from the mother's blood pass to the baby. This type of immunity lasts for only a few months until the baby's own immune system can function.

This chart depicts the number of vaccines that have been developed over the last two centuries. The first was in 1796 by Edward Jenner when he successfully vaccinated a child against smallpox. Then no vaccines were developed for 87 years. During the twentieth century, greater knowledge of pathogens and diseases led to the development of more vaccines.

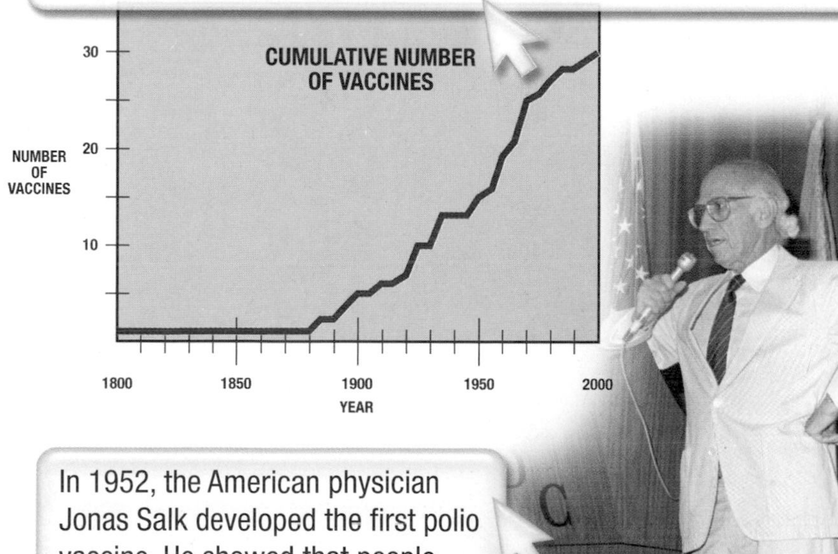

In 1952, the American physician Jonas Salk developed the first polio vaccine. He showed that people injected with dead polio viruses did not get the disease because they had developed antibodies against it.

LINKS

Technology Link
Prepare a special medical brochure for a foreign country that has travel alerts because of an infectious disease. Use the Internet to research this information from the World Health Organization website. Highlight the disease, its causes, preventive vaccines or medicine to take before travel, and treatments for the disease. Include drawings or colored photos related to the disease and its pathogen.

History Link
Research the history of how science has progressed in fighting infectious disease. Prepare a time line that includes at least six significant events. Give a presentation of the time line to your classmates.

Language Arts Link
Interview a doctor who diagnoses and treats infectious diseases. Prepare at least five questions to ask about the symptoms of two infectious diseases, their treatments, and whether or not the body builds immunity to the diseases. Write a brief report about the interview and present it to the class.

GLOSSARY

abyssal plain a large, nearly flat region on the ocean floor (page 108)

acceleration the change in an object's velocity over a period of time (page 78)

acid precipitation the precipitation that is more acidic than normal rainwater (page 12)

acne a skin condition that results when mostly excess sebum clogs skin pores (page 156)

active immunity a type of immunity in which the immune system produces antibodies in response to the presence of a pathogen (page 174)

acute beginning suddenly and lasting for a short time (page 168)

adolescence the period of time between childhood and adulthood (page 152)

air mass a very large body of air that maintains uniform temperature, humidity, and pressure as it moves (page 132)

alevin a very young salmon hatchling (page 22)

allergen a substance that provokes allergies (page 170)

allergy an abnormal reaction to a foreign substance (page 170)

B

biogeochemical cycle the cycling of chemical elements through the living and nonliving parts of an ecosystem (page 2)

biomass the organic matter from plant material and animal waste used as a source of energy (page 124)

botanist a scientist who studies plant life (page 42)

buoyant force an upward force in fluids that opposes gravity (page 62)

cancer a disease caused by cells that go through uncontrolled cell division (page 36)

cavity an area of tooth decay caused by prolonged exposure to bacteria (page 160)

cell cycle a series of events occurring during the life of a cell (page 32)

cell membrane a flexible structure that protects and controls what goes in and out of the cell (page 28)

cellular respiration the process in cells by which oxygen and glucose are used to produce energy and carbon dioxide (page 8)

cell wall the rigid outer layer that protects and supports plant cells (page 28)

centi- a prefix meaning one-hundredth of a unit (page 54)

charge the property of a particle that causes it to attract or repel other particles (page 90)

chemical bond the force of attraction that holds atoms together within a molecule (page 66)

chemical change an event that rearranges chemical bonds and forms one or more new substances with different properties (page 70)

chemical property a characteristic that describes when or how a substance will interact with other substances (page 70)

chromosome a structure located in the nucleus that contains DNA (page 30)

chronic lasting for a long time and/or recurring frequently (page 168)

circuit a complete path through which an electric current can flow (page 92)

coal a solid fossil fuel formed from decomposed plant remains (page 118)

collection the process by which precipitation is gathered into a body of water (page 6)

colonize to establish growth in an ecological community (page 40)

combustibility the ability to burn (page 70)

compost a mixture that consists largely of decayed organic matter and is used to fertilize soil (page 122)

Coriolis effect the way Earth's rotation affects wind and ocean currents (page 110)

cotyledon the nutrient-rich structure in a seed (page 20)

crescent the moon phase in which less than half of the moon's sunlit side is visible (page 144)

current electricity the continuous flow of electric charge through a material (page 92)

cytokinesis the final stage of the cell cycle in which the cytoplasm divides (page 32)

cytoplasm the jelly-like substance found inside the cell that contains organelles (page 28)

D

deci- a prefix meaning one-tenth of a unit (page 54)

deep ocean current a stream of water that flows deep beneath the ocean's surface (page 110)

denitrification the process that releases nitrogen from soil back into the atmosphere (page 10)

density the measure of how compact matter is, where mass is compared to volume (page 60)

desalination the process of removing salt from seawater to obtain freshwater (page 116)

disease any abnormal condition that causes illness in a living organism (page 164)

displace to push aside (page 62)

DNA [deoxyribonucleic acid] the molecule that contains the genetic material of a cell (page 30)

doldrums a relatively calm area near the equator where warm air rises (page 130)

E

ecological succession the predictable and orderly changes that occur over time in a community of plants and animals (page 38)

ecology the scientific study of the relationships between organisms and their environments (page 40)

effort the force that is applied to a simple machine (page 84)

electromagnet a magnet that produces a magnetic field by means of an electric current (page 96)

El Niño a climate event in the Pacific Ocean caused by wind shifts, resulting in unusually warm water temperatures (page 110)

embryo an organism produced from a spore or fertilized egg, in the early stages of development (page 20)

emission a pollutant released into the atmosphere (page 12)

endocrine system a network of glands that produces and releases hormones (page 154)

energy the ability to do work (page 80)

equinox the two days of the year when day and night are almost the same length everywhere on Earth (page 142)

extrusive a type of igneous rock that cools quickly and has small crystals (page 104)

F

ferromagnetic relating to matter with strong magnetic properties (page 96)

fertilization the uniting of the genetic material of two cells (page 20)

fluid a substance that can flow (page 62)

fossil fuel a source of energy formed from dead plants and animals (page 8)

front the boundary at which two different air masses meet (page 132)

fruit a plant structure that surrounds the seed(s) of a flowering plant (page 20)

fulcrum the point where a lever rotates or pivots (page 82)

fungus a decomposer that usually reproduces through spores (page 18)

gene a segment of DNA located on the chromosome that, among other things, controls specific traits (page 30)

generation one complete life cycle (page 14)

generator a machine that produces electricity from moving parts (page 98)

genetic disorder an abnormal condition that is usually inherited through genes (page 170)

genetic material the set of instructions within a cell that controls an organism's characteristics and life processes (page 14)

geothermal energy the energy produced by heat trapped within the earth (page 124)

gibbous the moon phase in which more than half, but not all, of the moon's sunlit side is visible (page 144)

gills the structures on the underside of a mushroom (page 18)

gland a specialized tissue that produces and releases chemicals (page 154)

grade a measure of the steepness of a slope (page 86)

H

hardwood the wood from broad-leaved, mostly deciduous trees (page 44)

homeostasis a state of balance within a cell, organ, or system (page 154)

hormone a chemical messenger that creates a response in the body (page 154)

horse latitudes the relatively calm areas located near both 30° latitudes where cool air sinks (page 130)

hydrocarbon a compound made of the elements hydrogen and carbon (page 118)

I

immune response the series of reactions that take place when white blood cells fight against infection (page 172)

immunity the body's ability to protect itself from pathogens before they cause disease (page 174)

inclined plane a straight, slanted surface used to raise an object (page 82)

infectious disease a contagious disease caused by pathogens (page 166)

insoluble unable to be dissolved in a given solvent (page 68)

interphase the stage of cell growth occurring at the beginning of the cell cycle (page 32)

intrusive a type of igneous rock that cools slowly and has large crystals (page 104)

isobar a line used on a map or chart connecting points of equal air pressure (page 134)

joule a metric unit used to measure work, equal to one newton meter (Nm) (page 80)

kilo- a prefix meaning one thousand units (page 54)

L

land breeze a local wind that blows from the land toward a body of water (page 128)

landfill a site for the disposal of waste materials (page 122)

lava the magma that reaches the surface of the earth (page 104)

law of universal gravitation every object in the universe attracts every other object, depending on their masses and distances from one another (page 140)

legume a plant that hosts nitrogen-fixing bacteria in nodules on its roots (page 10)

lever a beam or bar that pivots on a fixed point (page 82)

life cycle the series of stages that an organism passes through from egg or spore to reproducing adult (page 14)

lithosphere the crust and solid, upper portion of the mantle (page 106)

load an object that resists the force applied by a simple machine (page 84)

long bone one of several elongated parts of the human skeleton (page 158)

lunar eclipse an event that occurs when Earth passes directly between the sun and the moon, causing Earth's shadow to block the sun's light from the moon (page 144)

lymphatic system a major component of the immune system that protects the body from pathogens (page 172)

lymph node a rounded mass of tissue that filters fluids from other body tissues and traps pathogens (page 172)

M

mare a dark, flat area on the moon's surface, filled with hardened lava (page 148)

mass a measure of the amount of matter of an object (page 58)

metal a moldable substance that can reflect light and conduct heat and electricity (page 120)

milli- a prefix meaning one-thousandth of a unit (page 54)

mitosis the stage of the cell cycle in which the cell's nucleus divides in two (page 32)

mixture a combination of two or more substances that can be physically separated (page 68)

monsoon a wind that changes direction with the change of seasons and which usually brings very heavy rainfall (page 136)

mushroom a fleshy, spore-producing growth of certain fungi (pagc 18)

mutation a change in the DNA sequence of a gene or chromosome (page 36)

natural gas a mixture of methane and other gases formed from decomposed marine organisms (page 118)

natural resource a material found in nature that is useful to humans (page 114)

neap tide a condition of the least difference between low and high tides (page 146)

newton a metric unit used to measure force (page 80)

nitrogen fixation the conversion of nitrogen in the atmosphere to a form that can be used by plants (page 10)

nodule a swelling on a plant root that contains nitrogen-fixing bacteria (page 10)

noninfectious disease a disease that is not caused by pathogens and is not contagious (page 166)

nonrenewable not capable of being replenished or replaced within a sufficient period (page 116)

nuclear change an event that greatly alters the nucleus of an atom (page 72)

O

ore a naturally occurring rock from which useful metals or other minerals can be extracted (page 120)

organelle one of several tiny structures within a cell (page 28)

parallel circuit a circuit with more than one path for an electric current (page 94)

parr a young, growing salmon (page 22)

pathogen an organism or particle that causes disease (page 166)

perspiration a fluid released by the sweat glands (page 156)

petroleum a crude oil, which is a liquid fossil fuel formed from microscopic plants, animals, and marine organisms (page 118)

photochemical smog the brown smog produced when air pollutants react with sunlight (page 12)

photosynthesis the process that allows green plants to make sugars from sunlight, water, and carbon dioxide, and that releases oxygen into the atmosphere (page 8)

physical change an event that alters the form or size of matter but does not change the type of matter (page 64)

physical property a characteristic of matter that can be observed or measured without changing the type of matter (page 64)

pioneer species the first species to populate or repopulate a barren or disturbed area (page 40)

pituitary gland a gland in the brain that stimulates growth and development (page 154)

polar easterlies the cold winds from the poles to the 60° latitudes that blow from east to west (page 130)

polarity the quality of having two opposite poles—one positive and one negative (page 112)

primary succession the series of changes occurring in a newly formed, barren habitat (page 40)

puberty the process of physical changes as a child's body develops into an adult (page 152)

redd a salmon nest (page 22)

renewable capable of being replenished or replaced (page 116)

resistance a measure of how difficult it is for an electric current to flow (page 94)

runoff the precipitation that flows over the surface of soil (page 6)

S

sea breeze a local wind that blows from an ocean or sea toward the land (page 128)

seamount an undersea volcanic mountain found on the ocean floor (page 108)

sebum an oil produced by glands located in the skin (page 156)

secondary succession the series of changes occurring in an area where the existing ecosystem has been disturbed (page 44)

seedling a young plant (page 20)

sere a stage in the series of ecological succession (page 42)

series circuit a circuit with a single path for an electric current (page 94)

simple machine a device that makes work easier and which has few or no moving parts (page 82)

smolt an immature salmon that migrates to the ocean (page 22)

softwood the wood from conifers or evergreen trees (page 44)

solar eclipse an event that occurs when the moon passes directly between the sun and Earth, causing the moon's shadow to block the sun's light from a portion of Earth (page 144)

solstice the two days of the year in which the sun's most direct rays reach farthest north or farthest south (page 142)

soluble able to be dissolved in a given solvent (page 68)

solution a type of mixture of two or more substances that are evenly distributed (page 68)

speed a measure of the distance an object moves in a given amount of time (page 78)

spring tide a condition of the greatest difference between low and high tides (page 146)

standard unit an established quantity used for measurement and comparison (page 52)

static discharge a sudden flow of static electricity from one object to another (page 90)

static electricity the buildup of nonflowing electric charge (page 90)

subduction the process by which one lithospheric plate is forced under another (page 108)

substance a single kind of matter that is pure and possesses a specific set of properties (page 64)

substrate the base, usually rock or soil, in which an organism lives (page 40)

tide the periodic rising and falling of the surface level of ocean water (page 146)

topography the physical surface features of a place or region (page 48)

trade winds the winds that blow easterly from both 30° latitudes toward the equator (page 130)

trait a characteristic that can be passed on to an organism's offspring (page 30)

upwelling the movement of cold, deep ocean water to the ocean's surface (page 110)

vegetation the plant life in an area (page 42)

velocity a measure of an object's speed and direction (page 78)

virus a tiny, nonliving particle that can infect cells and cause disease (page 168)

voltage the power of an electric current, measured in volts (page 92)

volume the amount of space matter occupies (page 56)

W

wane to shrink (page 144)

wax to grow (page 144)

weather the condition of the atmosphere at any given moment in a particular area (page 126)

weight a measure of the pull of gravity on an object (page 58)

westerlies the winds, found between the 30° and 60° latitudes, that blow toward the poles from west to east (page 130)

wind the movement of air caused by differences in air pressure (page 126)

wisdom teeth the third molars, usually the last teeth to appear (page 160)

work the result of a force applied to an object that causes it to move (page 80)

INDEX

Boldfaced page numbers indicate the location of vocabulary word definitions.

E

F

G

CREDITS

Chapter 1

Snowpack sampling, **page 6**, California Department of Water Resources

Mississippi watershed, **page 7**, Jon Platek

Volvox, **page 8**, Jason K. Oyadomari, www.keweenawalgae.mtu.edu

Nodules (Nódulos con bacterias simbiontes (Rhizobium) en raíces de leguminosa), **page 10**, Laura Arribas

Calcium, **page 13**, Ellen Denny/Hubbard Brook Ecosystem Study

Chapter 2

Frog (*Litoria caerulea*), **page 14**, copyright by David A. de Groot, 2006

C. elegans, **page 15**, © Dennis Kunkel Microscopy, Inc.

Tadpole, **page 15**, courtesy of Lydia Fucsko at Lydia Fucsko frogs.org.au

Escherichia coli, **page 17**, © Dennis Kunkel Microscopy, Inc.

Death Cap, **page 18**, Dr. Robert Thomas Orr, © California Academy of Sciences

Salmon eggs, parr, and smolt, **page 22**, E. Peter Steenstra/U.S. Fish and Wildlife Service

Salmon alevin, **page 22**, © Michael Durham

Plasmodium falciparum, **page 24**, Steven Glenn, Laboratory and Consultation Division/CDC

Mosquito, **page 24**, CDC

Chapter 3

Cell mitosis, **page 33**, W. J. Moore/University of Dundee/Wellcome Images

Skin cancer, **page 37**, Dermatology Branch, National Cancer Institute

Lung cancer X-ray, **page 37**, National Cancer Institute

Chapter 4

Secretary of State William Seward, **page 42**, National Archives

Erratic, **page 42**, USGS

Black crust, **page 42**, NPS, Neal Herbert

Rock flour, **page 43**, USGS

Alder, **page 43**, Arley Muth/Corps of Engineers Regulatory

Dryas, **page 43**, U.S. Fish and Wildlife Service

Cottonwood, **page 43**, H. W. Phillips/Montana Fish, Wildlife, and Parks

Firefighters and burned forest, **page 46**, Jim Peaco/ National Park Service

Ground fire, **page 46**, National Park Service

Pinecone, **page 47**, Don Despain/National Park Service

Grasses and wildflowers; conifers, **page 47**, Jim Peaco/ National Park Service

Forest growth, **page 47**, Steven Erat

Snow-covered Mount Saint Helens, **page 48**, Jim Nieland/U.S. Forest Service

Post-volcanic Mount Saint Helens (in circle), **page 48**, Harry Glicken/ USGS

Trees and fireweed, **page 48**, Lyn Topinka/USGS/ Cascades Volcano Observatory

Valdez spill, **page 49**, *Exxon Valdez* Oil Spill Trustee Council

Oiled bird, **page 49**, Igor Golubenkov

Deepwater Horizon oil rig, **page 49**, USCG

Chapter 5

Anaconda: Judah Epstein and local villagers caught a live 15-foot anaconda in the Los Llanos region of the Venezuelan Amazon; **pages 52–53**; photo taken by Ryan Gaston.

Mars Climate Orbiter, **page 55**, Corby J. Waste/NASA/ JPL-Caltech

Astronaut, **page 58**, NASA

Iridium meteorite, **page 60**, NASA/JPL-Caltech/ Cornell

Astronaut in Neutral Buoyancy Lab, **page 62**, NASA

Mars, **page 63**, NASA

Chapter 6

Chlorine, **page 71**, Ben Mills

Stars, **page 74**, NASA

Shuttle launch, **page 74**, NASA

NOS kit, **page 75**, Holley Performance Products

Chapter 7

Shuttle, **page 77**, NASA

Sir Isaac Newton, **page 80**, Mezzotint by James McArdell after Enoch Seeman, 1760, Library of Congress

Jaws of Life, **page 85**, US Navy Journalist 1st Class Ralph Radford

Masada, **page 86**, Gary R. Gordon

Chapter 8

Statue, **page 92**, J. Patrick Fischer

Magnetite, **page 96**, Stan Celestian/Glendale Community College of Arizona

Wind farm, **page 98**, Michael and Cheryl Chiapperino

Chapter 9

DART buoy, **page 107**, NOAA

DART location map, **page 107**, NOAA

Ship using sonar, **page 108**, NOAA

Mauna Kea, **page 108**, Vadim Kurland

HMS *Challenger*, **page 109**, NOAA

Gulf Stream, **page 110**, NASA/Gulf Stream Tutorial

Jason 1, **page 111**, NASA-JPL

JOIDES Resolution, **page 112**, IODP/TAMU

Chapter 10

Rock salt, **pages 114 and 120**, Sundaram Overseas Operation, India, www.soo.co.in

Fluorescent minerals, **page 120**. This image is the property of Hannes Grobe and is used according to the stipulations given at http://creativecommons.org/licenses/by-sa/2.5. Neither Hannes Grove nor Creative Commons endorses, sponsors, or authorizes this textbook.

Crocoite, **page 121**, Andrew Silver/BYU Mineral Specimens 441/USGS

Chapter 11

Circumzenithal arc, **pages 127 and 133**, Andrew G. Sarras; award-winning photographer, fine artist, and sculptor; www.saffasart.com

Rawinsonde, **page 129**, Peter Guest/U.S. Navy

Chapter 12

Color-enhanced moon, **pages 139 and 143**, NASA/JPL-Caltech

GRACE satellites, **page 140**, NASA/JPL-Caltech

Earth's gravity, **page 141**, provided by University of Texas Center for Space Research and NASA

Earth, **page 148**, NASA

Astronaut, **page 148**, NASA

Neil Armstrong, **page 148**, NASA

Moon rock, **page 149**, NASA-JSC

Moon surface, **page 149**, NASA

Chapter 13

Eyelash, **pages 153 and 159**, © Dennis Kunkel Microscopy, Inc.

Chapter 14

Oral microbes, **pages 164–165**, © Dennis Kunkel Microscopy, Inc.

World Health Organization, **page 165**, © P. Virot/WHO

Staphylococcus aureus, **page 166**, Janice Haney Carr and Jeff Hageman, M.H.S./CDC

Influenza virus particle, **page 167**, CDC/Dr. Erskine L. Palmer; Dr. M.L. Martin

Trichophyton rubrum, **page 167**, CDC/Dr. Libero Ajello

Parasitic amoeba, **page 167**, CDC/DPDx–Melanie Moser

Streptococci, **page 168**, CDC

Strep throat, **page 168**, Dr. Heinz F. Eichenwald/CDC

Tick, **page 168**, Dr. Amanda Loftis, Dr. William Nicholson, Dr. Will Reeves, Dr. Chris Paddock/CDC

Prion, **page 169**, Rocky Mountain Laboratories, National Institute of Allergy and Infectious Diseases

Monocyte, **page 173**, © Dennis Kunkel Microscopy, Inc.

T and B cells, **page 173**, © Dennis Kunkel Microscopy, Inc.

Jonas Salk, **page 175**, CDC